Hydrodynamics of Estuaries

Volume II
Estuarine Case Studies

Editor
Björn Kjerfve, Ph.D.
Professor
Marine Science Program
Department of Geology
Belle W. Baruch Institute for
Marine Biology and Coastal Research
University of South Carolina
Columbia, South Carolina

CRC

CRC Press, Inc.
Boca Raton, Florida

Library of Congress Cataloging-in-Publication Data

Hydrodynamics of estuaries.

Bibliography: p.
Includes indexes.
1. Estuarine oceanography. 2. Hydrodynamics.
I. Kjerfve, Björn, 1944- .
GC97.H93 1988 551.46'09 87-21856
ISBN 0-8493-4369-0 (set)
ISBN 0-8493-4370-4 (v. 1)
ISBN 0-8493-4371-2 (v. 2)

Direct all inquiries to CRC Press, Inc., 2000 Corporate Blvd., N.W., Boca Raton, Florida, 33431.

International Standard Book Number 0-8493-4369-0 (set)
International Standard Book Number 0-8493-4370-4 (v. 1)
International Standard Book Number 0-8493-4371-2 (v. 2)

Library of Congress Number 87-21856
Printed in the United States

INTRODUCTION

Books on various aspects of physical oceanography abound. However, they are usually biased toward "blue water oceanography", emphasizing the synoptic aspects of oceanic processes away from continental landmasses. Whereas some do, others do not pay attention to dynamic processes on the continental shelf. Most ignore the physics of estuaries, lagoons, and bays, where the boundary conditions are likely to be highly complex, and numerical rather than analytical solutions are required to simulate hydrodynamic and dispersive characteristics.

The first book[1] exclusively devoted to estuarine characteristics and processes is a multi-authored text, which equally treats biological, chemical, geological, and physical aspects of estuaries. In addition, there exist a number of advanced texts that exclusively deal with physical characteristics and dynamics of estuaries. Two early books[2,3] do an excellent job of synthesizing salient hydrographic and dynamic features of estuaries from the point of view of an oceanographer. Similarly, with an oceanography bias, a handbook[4] of hydrography and sedimentation gives a practical how-to approach to the study of estuaries. Other early treatises[5-7] emphasize the engineering/sediment aspects of estuaries, and the engineering emphasis is similarly evident in two more recent estuarine books.[8,9] Over the past decade, a crop of multi-authored estuarine books with a physical slant have appeared,[12-12] having resulted from symposia/workshops. Although very up to date at the time of publication, they suffer somewhat from a coherent theme. In addition to Lauff's[1] early estuarine summary, numerous volumes have been devoted to the estuarine environment and contain one or more chapters on hydrodynamic or physical aspects of estuaries. These volumes are too numerous to name. However, the Estuarine Research Federation publishes on a recurring basis the proceedings from its biennial symposium,[13-19] and these volumes contain many useful articles on the hydrodynamics of estuaries. Finally, the National Academy of Sciences[20,21] has published two useful volumes that identify estuarine problem areas and research priorities.

The present volumes are an attempt to summarize many of the prevalent concepts and approaches in the investigation into hydrodynamics and physical processes of estuaries. It is difficult to find a current account of estuarine physical oceanography in any one place in the literature. It is my hope that *Hydrodynamics of Estuaries* will help fill this void. By having asked leading scientists in the field to contribute chapters, which are broadly summarizing in nature, I am hoping that these volumes will prove useful to oceanography students, research workers in the field, and to persons charged with the management of our estuarine resource.

These books are divided into two volumes. The first one focuses on estuarine physics and physical processes and interpretations. I have, for most parts, intentionally downplayed engineering applications to estuaries. It is my bias that a deeper understanding is accomplished with a physical approach, whereas an engineering approach is largely geared toward finding a solution to a problem. Of course, it is not always easy to make this distinction. The second volume is a presentation of physical case studies of several important estuaries, spanning the major geomorphic types. I believe that it can be very useful to have such information gathered in one place. Without attempting to give equal play to all areas of the world, I have consciously strived to be more international in scope in selection of both authors and estuarine case studies.

I would like to extend my thanks to the chapter authors for being so patient with me. To edit this volume required a much larger time commitment than I initially envisioned, and the authors certainly had reason to be impatient with me for taking so long. Also, I would like to thank those scientists who provided their valuable time to review the manuscripts. Finally, I would like to than K. E. Magill of the Belle W. Baruch Institute for Marine Biology and Coastal Research, University of South Carolina, for invaluable help with English editing of the manuscripts and handling editorial matters during my many international journeys during the course of book preparation.

REFERENCES

1. **Lauff, H. G., Ed.,** *Estuaries,* Publication 83, American Association for the Advancement of Science, Washingtion, D.C., 1967.
2. **Dyer, K. R.,** *Estuaries: A Physical Introduction,* John Wiley & Sons, London, 1973, 140.
3. **Officer, C. B.,** *Physical Oceanography of Estuaries and Associated Coastal Waters,* John Wiley & Sons, New York, 1976, 465.
4. **Dyer, K. R., Ed.,** *Estuarine Hydrography and Sedimentation,* Cambridge University Press, Cambridge, 1979, 230.
5. **Ippen, A. T., Ed.,** *Estuary and Coastline Hydrodynamics,* McGraw-Hill, New York, 1966, 744.
6. **Bruun, P.,** *Stability of Tidal Inlets,* Elsevier, Amsterdam, 1978, 510.
7. **McDowell, D. M. and O'Connor, B. A.,** *Hydraulic Behaviour of Estuaries,* John Wiley & Sons, 1977, 292.
8. **Fischer, H. B., List, E. J., Koh, R. C. Y., Imberger, J., and Brooks, N. H.,** *Mixing in Inland and Coastal Waters,* Academic Press, New York, 1979, 483.
9. **Fischer, H. B., Ed.,** *Transport Models for Inland and Coastal Waters,* Academic Press, New York, 1981, 542.
10. **Kjerfve, B., Ed.,** *Estuarine Transport Processes,* University of South Carolina Press, Columbia, 1978, 331.
11. **Sundermann, J. and Holz, K.-P., Eds.,** *Mathematical Modelling of Estuarine Physics,* Springer-Verlag, Berlin, 1980, 265.
12. **van de Kreeke, J., Ed.,** *Physics of Shallow Estuaries and Bays,* Springer-Verlag, Berlin, 1986, 280.
13. **Cronin, L. E., Ed.,** *Estuarine Research,* Vols. 1 and 2, Academic Press, New York, 1975, 738 and 587.
14. **Wiley, M., Ed.,** *Estuarine Processes,* Vols. 1 and 2, Academic Press, New York, 1976, 541 and 428.
15. **Wiley, M. L., Ed.,** *Estuarine Interactions,* Academic Press, New York, 1978, 603.
16. **Kennedy, V. S., Ed.,** *Estuarine Perspectives,* Academic Press, New York, 1980, 533.
17. **Kennedy, V. S., Ed.,** *Estuarine Comparisons,* Academic Press, New York, 1982, 709.
18. **Kennedy, V. S., Ed.,** *The Estuary as a Filter,* Academic Press, New York, 1984, 511.
19. **Wolfe, D. A., Ed.,** *Estuarine Variability,* Academic Press, New York, in press.
20. **Geophysics Study Committee, Eds.,** *Estuaries, Geophysics and the Environment,* National Academy of Sciences, Washington, D.C., 1977, 127.
21. **Geophysics Study Committee,** *Fundamental Research on Estuaries: The Importance of an Interdisciplinary Approach,* National Academy of Sciences, Washington, D.C., 1983, 79.

THE EDITOR

Björn Kjerfve, Ph.D., was born in Skövde, Sweden and is currently Professor of Marine Sciences and Geological Sciences and a research associate with the Belle W. Baruch Institute for Marine Biology and Coastal Research at the University of South Carolina in Columbia.

Prof. Kjerfve received his B.A. in mathematics from Georgia Southern College in 1968, his M.S. in oceanography from the University of Washington in 1970, and his Ph.D. in marine sciences from Louisiana State University. He has been on the faculty at the University of South Carolina since 1973.

Prof. Kjerfve's current research includes transport and modeling studies in estuaries and coastal lagoons. In addition to his projects in the U.S., his research has taken him to sites in Europe, Australia, Southeast Asia, and Central and South America. He has published numerous scientific articles, technical reports, and is the editor of *Estuarine Transport Processes*. Prof. Kjerfve is currently on the editorial board of *Coral Reefs, Trabalhos Oceanograficos* (Recife, Brazil), *Anales del Instituto de Ciencias del Mar y Limnologia and Publicaciones Especiales* (Mexico, D. F., Mexico), and *Research Bulletin* (Phuket, Thailand). He is a member of the Estuarine Research Federation, American Geophysical Union, American Meteorological Society, Estuarine and Brackish Water Sciences Association (U.K.), and other organizations.

CONTRIBUTORS

Clifford A. Barnes, Ph.D.
Professor Emeritus
School of Oceanography
College of Oceans and Fisheries
University of Washington
Seattle, Washington

Harry H. Carter, Ph.D.
Professor Emeritus
Marine Sciences Research Center
State University of New York
Stony Brook, New York

R. H. Chapman
Electronic Engineer
Centre for Water Research
University of Western Australia
Nedlands, Western Australia

Curtis C. Ebbesmeyer
Vice President of Research
Evans-Hamilton, Inc.
Seattle, Washington

Mohammed I. El-Sabh, Ph.D.
Professor Doctor
Departement d'Oceanographie
Universite du Quebec a Rimouski
Rimouski, Quebec, Canada

Jorg Imberger, Ph.D.
Professor of Civil Engineering
Director of Centre for Water Research
University of Western Australia
Nedlands, Western Australia

Björn Kjerfve, Ph.D.
Professor
Department of Geology
Belle W. Baruch Institute
University of South Carolina
Columbia, South Carolina

Maynard M. Nichols, Ph.D.
Professor
School of Marine Science
College of William and Mary
Virginia Institute of Marine Science
Gloucester Point, Virginia

Donald W. Pritchard, Ph.D.
Professor and Associate Director for Research
Marine Science Research Center
State University of New York
Stony Brook, New York

Jean-Claude Salomon, Ph.D.
Professor
Oceanographie Physique
Universite d'Aix-Marseille
Marseille, France

William W. Schroeder, Ph.D.
Professor
Marine Science Program
University of Alabama
Dauphin Island, Alabama

Ned P. Smith
Senior Research Scientist
Department of Physical Oceanography
Harbor Branch Foundation, Inc.
Fort Pierce, Florida

William J. Wiseman, Jr., Ph.D.
Professor
Coastal Studies Institute
Louisiana State University
Baton Rouge, Louisiana

Eric Wolanski
Principal Research Scientist
Department of Physical Oceanography
Australian Institute of Marine Science
Queensland, Australia

Jack Q. Word
Battelle Northwest
Marine Research Laboratory
Sequim, Washington

TABLE OF CONTENTS

Chapter 10

OCEANOGRAPHY OF CHESAPEAKE BAY

Harry H. Carter and Donald W. Pritchard

TABLE OF CONTENTS

I. INTRODUCTION

The Chesapeake Bay estuarine system was formed by the most recent rise in sea level which began approximately 15,000 to 18,000 years ago with the retreat of the glaciers at the end of the Wisconsin glaciation. Sea level at that time was approximately 125 m below its present level. As sea level rose it advanced across the previously exposed continental shelf, reaching the present mouth of the Chesapeake Bay basin less than 10,000 years ago. As the sea rose it drowned the ancestral river valley system which had been carved during the previous low stand and transformed the riverine system into an estuarine system. The Chesapeake Bay estuarine system is shown in Figure 1.

The Chesapeake Bay is a dynamic environment characterized by marked variations in its chemical and physical properties on a wide range of space and time scales. These fluctuations may be produced by processes active within the Bay, or they may be the result of processes active at one end in the drainage basin, or at the other end in the ocean.

The purpose of this paper is to briefly summarize some of the more important features of the physical oceanography of Chesapeake Bay — the residual circulation and forcing mechanisms, the tides and tidal currents, and the salinity and temperature distributions.

II. TIDES AND TIDAL CURRENTS

Cotidal hours and cocurrent hours are defined in Schureman[1] as the average intervals in lunar hours between the moon's transit over the meridian of Greenwich and the time of the following high water and the time of maximum flood current*, respectively. Their distributions are shown in Figures 2 and 3. Corange lines, lines that pass through places of equal mean tidal range, are shown on Figure 4.

According to Hicks,[3] the observed tidal wave progresses from the entrance at the Capes to the mouth of the Susquehanna River at Havre de Grace in approximately 14 lunar hr. Since the tidal wave entering the Bay from the open ocean has a predominant period of 12 lunar hr, a crest does not quite traverse the length of the Bay before the next following crest enters at the Capes. In this respect, Chesapeake Bay is somewhat unique in that it is able to hold a complete semidiurnal tidal wave at all times.

The Bay is wide enough so that rotational effects are important. As a result, in the Bay below the Severn River, the tide has the characteristics of a Kelvin wave, with a slightly larger range on the eastern side than on the western side, and with maximum flood and maximum ebb occurring at nearly the same time as high water and low water, respectively. North of the Severn, friction and reflection result in characteristics which are intermediate between those of a pure progressive wave and those of a standing wave, but becoming asymptotic to the characteristics of a standing wave as one approaches the head of the Bay. Figures 2, 3, and 4 show this quite clearly; between the Capes and the mouth of the Severn River, high water and maximum flood are in phase, i.e., the wave is progressive but north of the Severn River, the time between maximum flood and the next high water increases. At Pooles Island, halfway between the mouth of the Severn River and Havre de Grace, maximum flood precedes high water by about 1 hr, and at Fishing Battery Light, 3 mi below Havre de Grace, high water lags maximum flood by about 2.6 hr, close to the 3.1-hr time difference characteristic of a standing wave. As a result, it is possible to sample at the same phase of the tide between the Capes and Pooles Island, some 18 mi south of the head of the Bay, if the sampling vessel proceeds northward at a speed of between 10 and 12 kn (~480 km/day). In the reach of the Bay north of Pooles Island, slack water occurs at substantially the same time at all locations.

* Average time of max flood, SBF + 3.1 hr, max ebb ± 6.1 hr, and SAF − 3.1 hr.

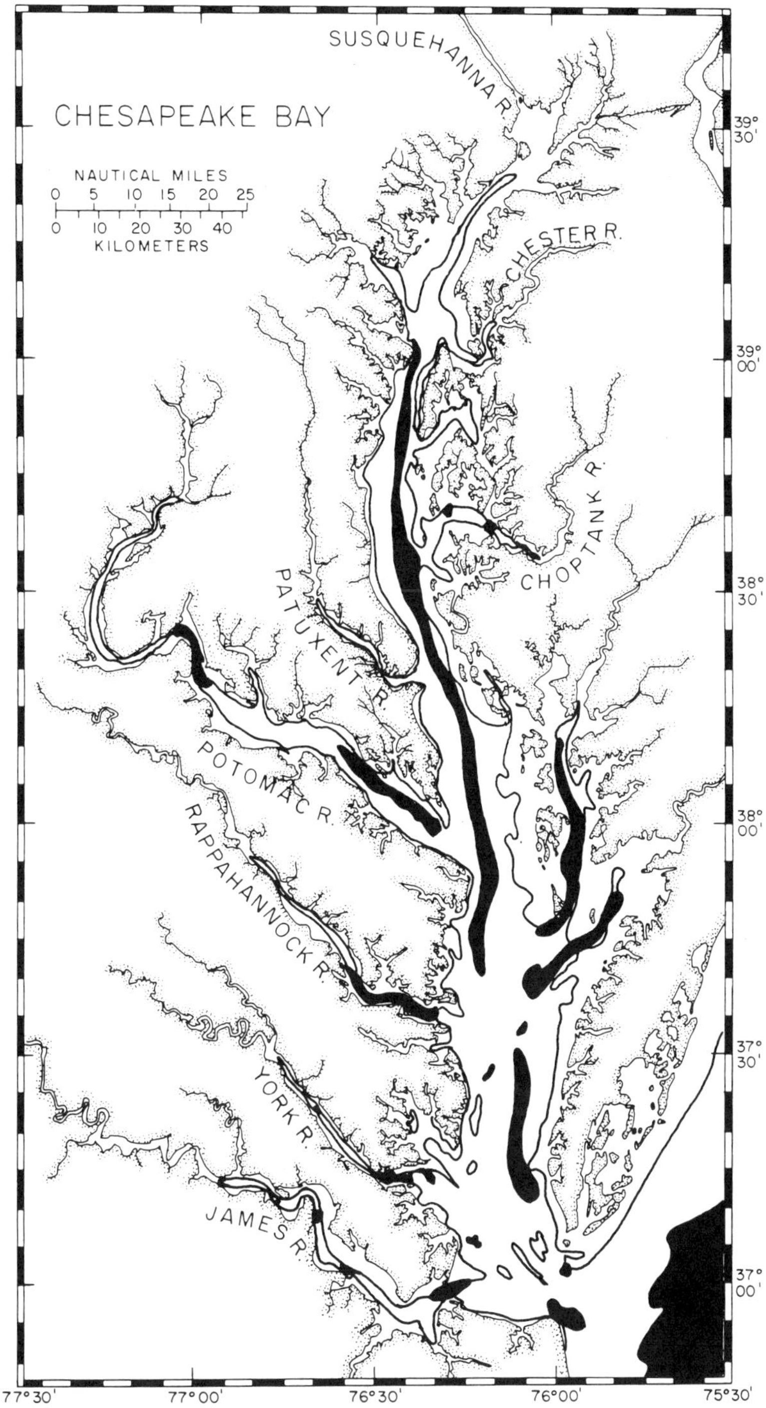

FIGURE 1. The Chesapeake Bay Estuarine System.[2]

The mean tidal range (Figure 4) decreases from about 3.0 ft (0.9 m) at the entrance to a minimum of about 1.0 ft (0.3 m) at Annapolis, then rises to 2.3 ft (0.7 m) at the head. The maximum range in the system is 3.9 ft (1.2 m) at Walkinton, Virginia on the Mattiponi River. As noted above, the range is significantly larger on the eastern shore of the Bay

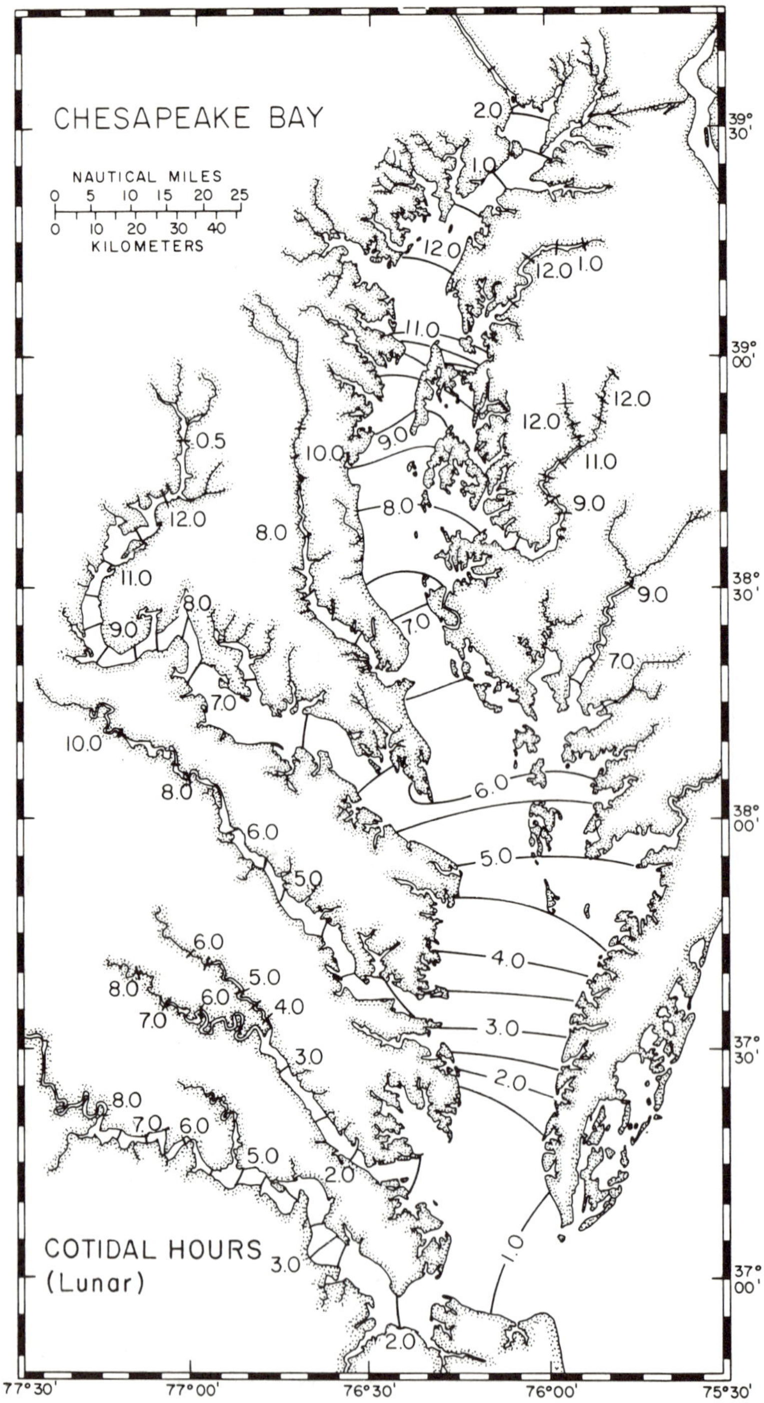

FIGURE 2. Cotidal hours (expressed in lunar hours).[3]

proper due to rotational effects. The difference is as large as 0.5 ft (0.15 m) and averages 0.2 ft (0.06 m). Velocities are also reported to be higher on the eastern side of the Bay as well,[4] though in part this may be due to the fact that the deep channel tends to be nearer the eastern shore, with a consequent lower frictional effect on that side of the Bay.

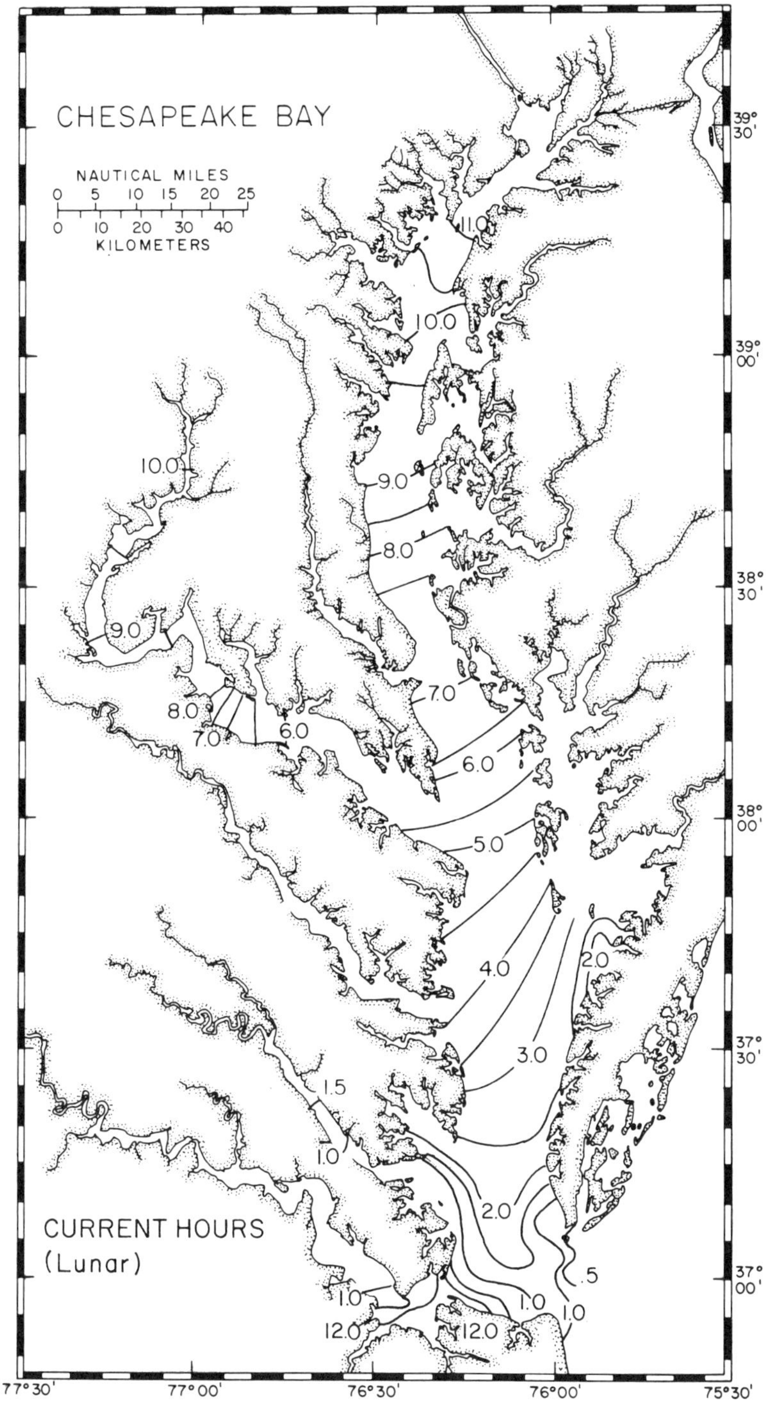

FIGURE 3. Current hours (expressed in lunar hours).[3]

Average tidal current amplitudes in the cross section at the mouth of the Bay vary from 1.25 kn (0.64 m/sec) to 2.00 kn (1.03 m/sec). The tidal current amplitudes decrease within the lower reaches of the Bay, such that at Wolf Trap Light, which is located between the mouth of the York River and the mouth of the Rappahannock River, the tidal current

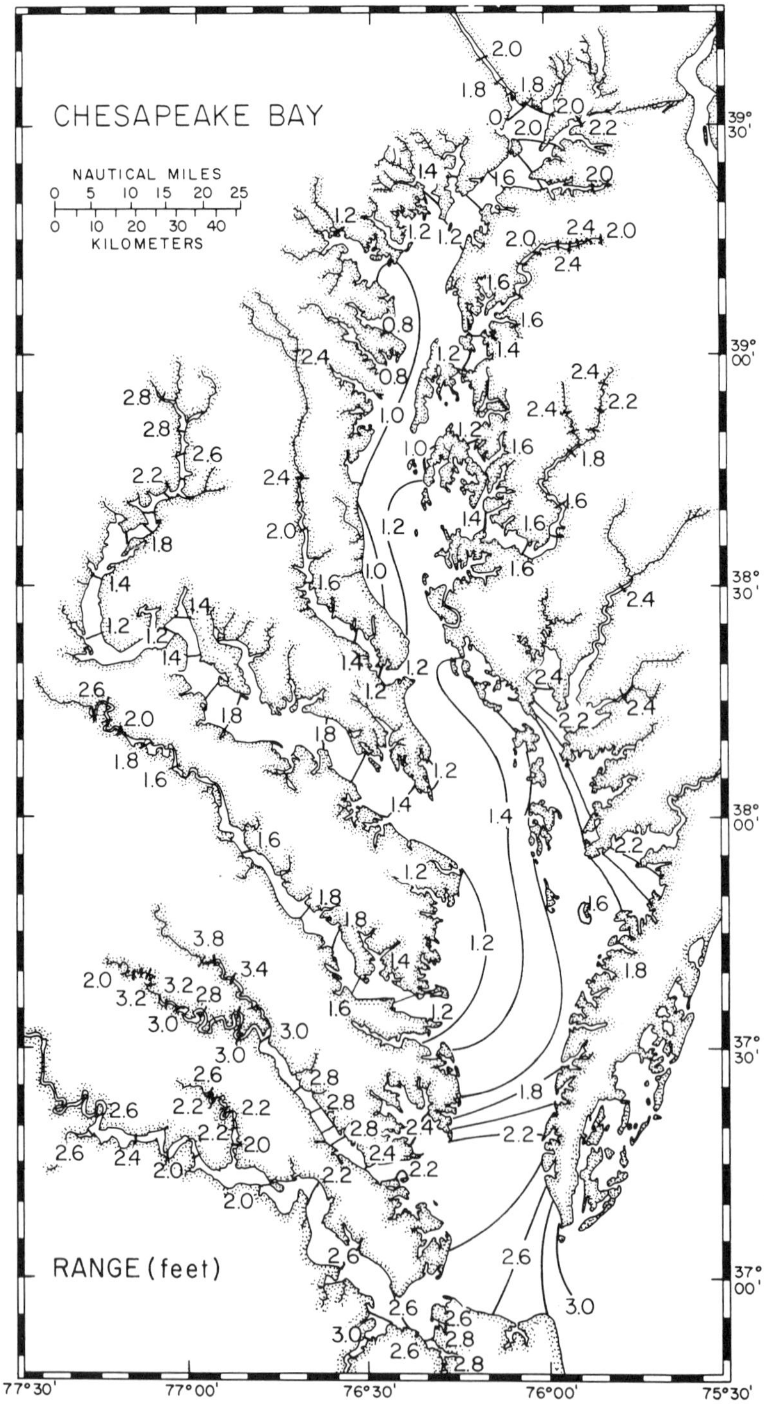

FIGURE 4. Mean tidal exchange in feet.[3]

amplitude is 1.10 kn (0.57 m/sec). Over the major portion of the middle reaches of the Bay the tidal current amplitudes range from 0.25 kn (0.13 m/sec) to 0.80 kn (0.41 m/sec), with the lower values occurring in the wider cross sections and the higher values in the narrower cross sections. The tidal current amplitude at the Chesapeake Bay Bridge is about 0.8 kn

Table 1[6]
CIRCULATION PATTERNS ACCORDING TO DIRECTION OF FLOW AT 3 LEVELS (10 M, 25 M, 40 M), PROBABILITY OF OCCURRENCE, AND DURATION

Type	Direction of flows	Frequency of occurrence	Mean duration (days)
Classical estuary	($U_{10} > 0$; $U_{40} < 0$; $U_{25} \lesseqgtr 0$)	43%	2.5
Reverse estuary	($U_{10} < 0$; $U_{40} > 0$; $U_{25} \lesseqgtr 0$)	21%	1.6
Three-layered	($U_{25} > 0$; $U_{10} < 0$; $U_{40} < 0$)	1	1.0
Reverse three-layered	($U_{25} < 0$; $U_{10} > 0$; $U_{40} > 0$)	7	1.5
Discharge	(All 3 positive)	6	1.3
Storage	(All 3 negative)	22%	1.6
		100%	

(0.41 m/sec), increasing to 1.15 kn (0.59 m/sec) off Worton Point (about 12 mi north of the entrance to Baltimore Harbor), and then decreasing to 0.55 kn (0.28 m/sec) at Fishing Battery Light on the Susquehanna Flats.

At springs (neaps) tidal current amplitudes are about 30 to 40% higher (lower) than the average values given in the previous paragraph.

III. RESIDUAL CURRENTS AND THEIR CAUSES

The residual circulation is that which remains after the tidal currents have been removed. Although the classical two-layered estuarine flow driven by the combination of the seaward sloping sea surface and the upstream or riverward longitudinal density gradient is the most frequent mode of residual flow, other components — primarily those driven by the near and far field winds — are major contributors at times. To illustrate, consider Elliott's[5,6] analysis of the records from a mooring consisting of three current meters at nominal depths of 10, 25, and 40 m which were maintained for a full year during 1974 to 1975 in the lower Potomac River estuary. The records were filtered to remove the major tidal components and the residuals averaged over each calendar day to produce daily estimates of the mean residual flows. An empirical orthogonal function (EOF) analysis in the time domain showed that:

1. Of the total velocity variance 79% could be explained by the first two modes.
2. The first mode (classical estuarine or reverse estuarine) was associated with wind forcing and contained 48% of the total velocity fluctuations. The classical pattern was associated with downstream wind, falling sea level locally, and a downstream surface slope; the mode was reversible.
3. The second mode (storage or discharge) contained 31% of the total velocity variance and was associated with rising (storage) or falling (discharge) sea level. This mode was considered to be due to nonlocal effects which were propagated into the Potomac from the Bay.

The frequencies of occurrence of the six patterns that emerged are summarized in Table 1. Although the classical two-layered flow pattern occurred for a surprisingly small 43% of the time it always emerged for averaging periods of 10 days or longer.

Analysis of a 3-month record from a two-current meter mooring established in the upper Chesapeake Bay near Howell Point[7] during the fall of 1975 also showed similar residual flow variability. This paper also showed that the variability was related to both local wind forcing which was driven by the component of the wind along the axis of the Bay, and to far-field wind forcing which appeared to be a result of Ekman transport produced by the cross-Bay component of the winds in the middle and lower reaches of the Bay.

Wang and Elliott examined the frequency response and the coupling between the Bay and the Potomac River with spectral and EOF analysis techniques. They used data collected during the first 2 months of Elliott's year long study consisting of wind, surface elevation, and bottom currents in the Potomac. Sea-level data from four locations in the Bay and the results of a 3-day intensive current measurement experiment in the Potomac[8] were also considered. Their analyses were reported in Wang and Elliott[9] and later in Elliott and Wang.[4]

From the records of wind and sea level they found the Bay wind and sea-level spectra to be markedly similar with peaks at 2.5, 5, and 20 days. The dominant sea-level fluctuations in the Bay had a period of 20 days and were considered to be the result of Ekman fluxes in the adjacent coastal ocean, i.e., water was driven out of (into) the Bay by northward (southward) winds. These fluctuations were then damped as they propagated up the Bay.

The 5-day fluctuations in Bay sea level had up-Bay phase propagation but with a slight increase in amplitude. In addition, Annapolis sea-level fluctuations were coherent with Patuxent east-west winds, all of which suggests response at this period to both coastal sea level and to local lateral winds (Ekman).

The 2.5-day oscillation in Bay sea level had down-Bay phase propagation with a marked decrease in amplitude between Annapolis and Kiptopeake Beach. They were coherent with north-south (longitudinal) winds at both Annapolis and Kiptopeake Beach and are considered, therefore, to be generated within the Bay by north-south winds. On the other hand, if the effective mean depth of the Bay is in fact the depth computed from the observed time of travel of the tidal wave from the mouth to the head of the bay (14 hr), then the longitudinal seiche period of the Bay is apparently close to 2.5 days. The fact that the 2.5-day oscillation shows an amplitude increase in the up-Bay direction is suggestive of the presence of longitudinal seiching.

In the Potomac River, surface slopes were not coherent with local longitudinal winds (northwest-southeast) but were coherent with surface slopes in the Bay, suggesting a coupled Bay-River response to north-south winds. The near-bottom currents in the River had spectral peaks at 20, 5, and 2.5 days.

For periods >10 days, the bottom currents were coherent with northeast-southwest (lateral) winds, suggesting a bottom flow compensation for a surface Ekman flux. They were not coherent with local sea level at these periods and thus were apparently unaffected by any 20-day fluctuations which propagated into the Bay from the coastal ocean. For periods <10 days (5 and 2.5 days), bottom currents were coherent with surface slope and less so with longitudinal wind stress in the River, which is suggestive of local forcing. However, as noted above, since Bay and River surface slopes were coherent, these currents were apparently, at least in part, nonlocally forced by longitudinal winds over the Bay. The bottom currents, however, were not coherent with sea level for time scales larger than 3 days. This suggests that the volume flux, i.e., rate of change of sea level, was largely confined to the upper part of the water column (second mode) and, of course, lagged sea level by 90° (15 hr). Since the bottom currents were driven by surface slope, they also led the volume flux, i.e., surface currents, by 90° (15 hr), indicative of an upward phase propagation at periods <3 days.

In summary, Elliott and Wang have shown that nonlocal forcing is an important mechanism for distributing energy between the coastal ocean, the Bay, and its tributaries. It is clear that observation programs >20 days are required if one is to educe all of the components of the residual circulation.

Another extensive set of current meter, water level, and wind records were collected in the fall of 1977 by investigators from the Chesapeake Bay Institute, just north of the mouth of the Patuxent River.[10] In that study, 22 current meters were deployed in two cross-bay sections, one extending easterly from Kenwood Beach on the western shore and the second located approximately 8 mi south of the first section extending easterly from Cove Point. Another 8 m were deployed on three separate moorings located on a diagonal between the two east-west sections. The meters were deployed between October 12 and November 2, 1977. Four tide gages were located at approximately the four corners of the study area.

A preliminary analysis of the data set was included in Pritchard and Rives.[10] More detailed analyses were made by Vieira,[11] Pritchard and Vieira,[12] and Vieira.[13] Vieira calculated and plotted the laterally integrated residual flux per meter of depth below MLW at the Kenwood Beach section between October 16th and 30th and found, like Elliott before, considerable variability in the residual circulation patterns. In 14 days, the circulation pattern changed eight times. Although the statistics are sparse, both occasions of discharge were followed by a period of classical estuarine flow and two of the three occasions of classical estuarine flow were followed by periods of storage, agreeing with Elliott's comments regarding flow sequences.[6] Averaged over the period of the experiment, however, the classical estuarine circulation pattern emerged.

Using the same data set (laterally integrated residual flux per meter of depth), Vieira removed the mean and trend to reduce the presence of gravitational effects and to highlight the shorter period, meteorologically driven, flows. These data, when plotted and compared to the wind stress calculated from winds from the nearby Patuxent Naval Air Station, showed the surface layers down to about 8 m responding directly to the up- and downbay component of the winds and propagating down the water column. Conversely, the bottom layers respond somewhat later (8 hr) in the opposite direction, presumably due to a downwind setup of the sea surface. This counter flow starts at the bottom and propagates upward reaching depths just below the pycnocline in about 20 hr. Calculation of the rotary coherence squared between slope of the sea surface and the wind stress showed significant coherence at all periods greater than about 3 days with maximums at 4 and 7 days.

Tributary estuaries to the Chesapeake Bay often display departures from the classical two-layered flow pattern that are not related to wind forcing. Several of the tributary estuaries on the western shore of the upper Bay have relatively small drainage areas, with a consequence that the freshwater which directly enters these tributaries contributes to the density driven residual flow pattern only in the upper reaches of these tidal waterways. The circulation patterns which occur over most of the length of these tributaries are controlled by processes occurring in the reach of the Bay adjacent to the mouths of the tributary estuaries. In those tributary estuaries which are shallow compared to the adjacent Bay, so as to be in communication only with the upper seaward flowing layer of the Bay, the following seasonal sequence occurs. As described by Pritchard,[14] high runoff from the Susquehanna during spring causes the salinity of the Bay waters to decrease, with the result that the salinity of the Bay waters become lower than the salinity in the adjacent shallow tributaries. There then occurs an inflow into the tributary at the surface and an outflow at the bottom. This reverse circulation extends well up into the tributary to the position where the maximum salinity in the tributary is found. This maximum in salinity is found halfway or more up these tributary estuaries for roughly half the year; an inverse estuarine flow pattern occurs below the maximum and a classical pattern occurs above this location.

During the summer and fall of the year the low flow from the Susquehanna causes the salinity in the Bay adjacent to these estuaries to increase, such that the salinity in the tributary is now less than the salinity in the adjacent Bay. During this period the classical two-layered flow pattern, with outflow at the surface and inflow at the bottom, prevails in these shallow tributary estuaries.

Baltimore Harbor is the archetype of a class of tributary estuaries having relatively small drainage areas with channel depths about equal to the channel depths in the adjacent parent estuary. As first described by Pritchard and Carpenter,[15] a three-layered flow pattern exists in Baltimore Harbor, with inflow from the adjacent Bay in the surface and bottom layers, and outflow from the Harbor in a mid-depth layer. This is a density driven flow pattern even though the vertically averaged density does not vary significantly along the length of this tributary waterway. A moderately strong stratification is maintained in the waters of the Chesapeake Bay adjacent to the mouth of Baltimore Harbor as a result of the counterflow of the low salinity upper layer and higher salinity deeper layer, in balance with only moderately intensive vertical mixing. Within the Harbor the counterflowing layers, which tend to maintain stratification against the effects of vertical mixing, are lacking and vertical stratification is weakened. In fact, vertical homogeneity usually exists at the inner end of the Harbor. Consequently, the surface waters of the Harbor are more saline than those in the adjacent Bay, while the bottom waters of the Harbor are less saline than the waters at the same depth in the adjacent Bay. The longitudinal density gradients at the surface and at the bottom therefore drive the waters in both of these layers into the Harbor; continuity requires that there be an outflow at mid-depth.

Owen[16] described what then appeared to be a unique departure from the classical estuarine circulation pattern in the Patuxent River. During a significant part of the year this estuary has a three-layered circulation pattern, similar to that characteristic of Baltimore Harbor, in the lower one third or so of the estuary, while in the upper two thirds or so a classical two-layered circulation pattern exists. At other times of the year the two-layered circulation pattern is found throughout the waterway.

Another cause of perturbations to the classical estuarine flow pattern has recently been identified.[17,18] They hypothesize that the advection of relatively freshwater into the mouth of the York River estuary from the Cheaspeake Bay during spring tides reduces, or even reverses, the normal pressure gradient which drives the gravitational circulation thus limiting the inflow of more saline bottom water and permitting the strong spring tidal currents to establish vertical homogeneity. When the tidal heights decrease, the process is reversed. This phenomena has also been found in other Virginia tributary estuaries to the Chesapeake Bay.

To this point, we have considered only the longitudinal component of the residual circulation. There are, of course, lateral and vertical components as well as lateral and vertical variations in the longitudinal component. Vertical flows and vertical variations in the longitudinal component of residual flow are, of course, essential parts of what we have earlier referred to as the gravitational or estuarine circulation.[19] These lateral variations in longitudinal residual flow and lateral residual flows are only now beginning to be studied in the Chesapeake Bay estuarine system.[20]

IV. SALINITY

According to Pritchard,[14] Bay salinity varies more or less regularly along the length of the Bay, from that of nearly full seawater at the mouth to that of the inflowing Susquehanna River water at the head of the Bay. The vertical distribution of salinity is characterized by an upper layer of very slow increase with depth, an intermediate layer of more rapid increase (the halocline), and a deep layer in which the salinity increase with depth is again small. The salinity also varies laterally across the Bay with lower salinities on the western side of the Bay. Although the greater runoff of freshwater from the western shore contributes to this difference, the major cause is the rotation of the earth.

Figures 5 through 8 show the characteristic features of the salinity distribution for each of the four seasons — winter, spring, summer, and autumn. For each season the horizontal distribution of salinity is given for the surface and the longitudinal-vertical distributions are shown as vertical sections taken along the axis of the Bay. Note that minimum salinities occur in spring, with essentially freshwater extending on the average to Pooles Island, and

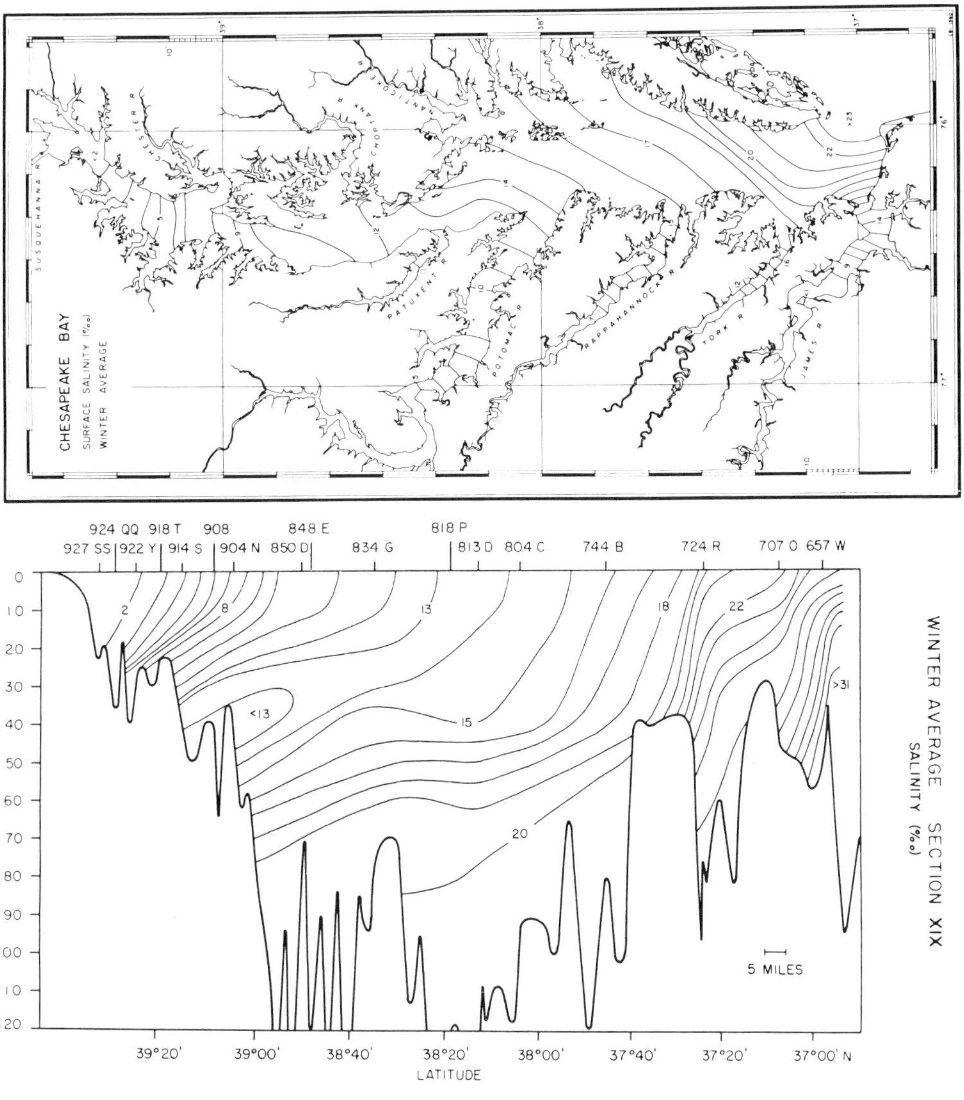

FIGURE 5. Horizontal distribution of average surface salinity in Chesapeake Bay — winter season.

maximum salinities in autumn, when low but measurable ocean-derived salt concentrations extend onto the Susquehanna Flats. Also note from the longitudinal-vertical sections that the mid-Bay between the Severn River and the York River is more stratified than the regions both riverward and oceanward. This has important implications for the mid-Bay lower layer dissolved oxygen concentrations in the summer.

There are also marked natural temporal salinity variations containing large monthly and interannual periods which are greatest in the upper reaches of the Bay and its tributary estuaries.[21]

To date, man has had little effect on the salinity distribution in the Bay or its tributaries. Changes in Bay salinity could result from flow regulation of the Susquehanna or enlargement of the C & D canal which connects the Bay with Delaware Bay. Pritchard[22] analyzed the effect of widening and deepening the canal during the early 1970s and concluded that the effect would be the greatest during periods of low river flow when salinities are a maximum and that the average maximum salinity would increase from 17.23 to 17.62‰ at the Bay Bridge, from 9.00 to 11.58‰ at Pooles Island, and from 2.14 to 2.94‰ at Turkey Point.

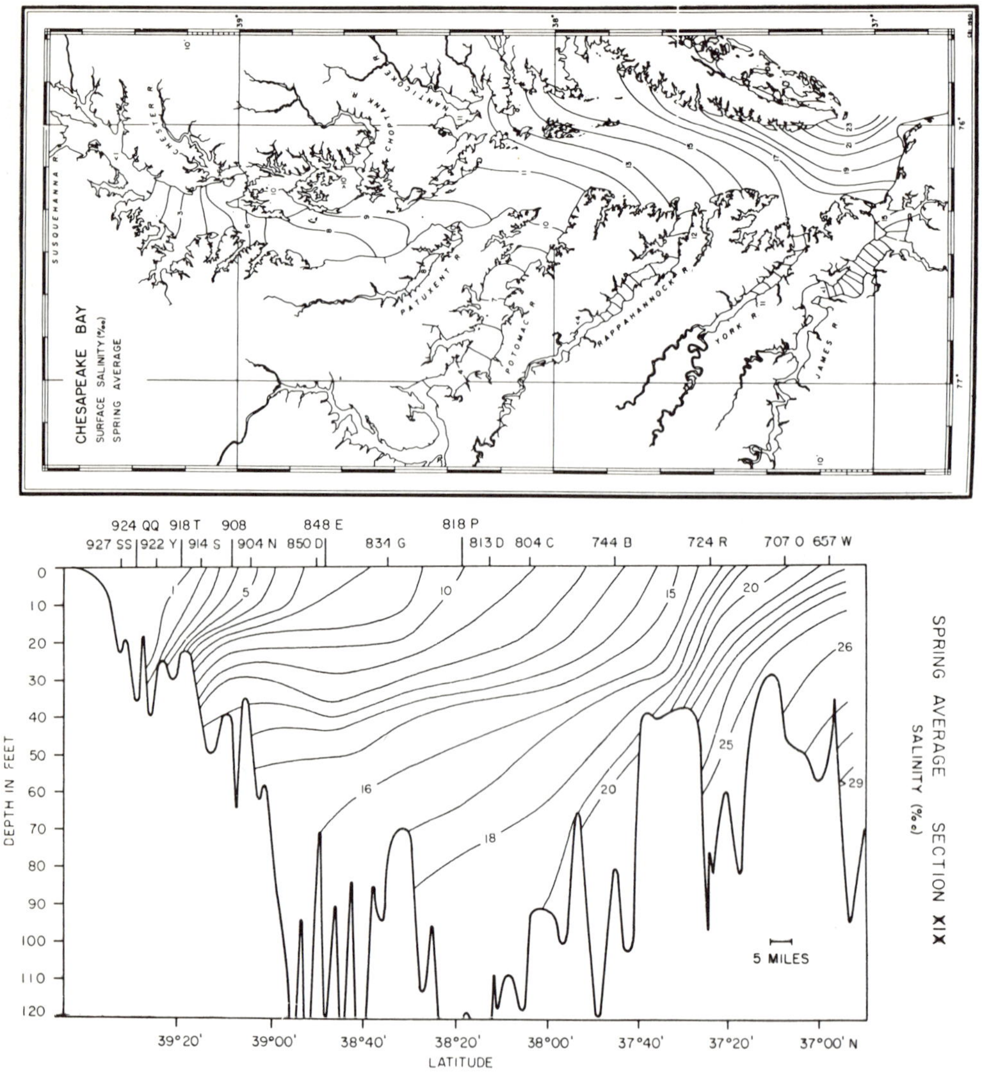

FIGURE 6. Horizontal distribution of average surface salinity in Chesapeake Bay — spring season.

V. TEMPERATURE

Temperature is an important oceanographic parameter in Chesapeake Bay because of its effect on density, on oxygen solubility, on a number of other physicochemical properties of seawater, and on biological activity. There are marked natural temporal and spatial variations of water temperature in the Chesapeake Bay system.

According to Seitz,[23] waters in Maryland are somewhat warmer than those in Virginia (0.5 to 3°C) during the summer months but the reverse is true during the balance of the year (October to May). On an annual basis, therefore, the average water temperatures in Virginia are about 0.5°C warmer than in Maryland.[21] Spatially, local gradients as high as 1°C/km are observed. Vertically, maximum top to bottom density differences due to temperature will be observed at a mid-Bay station during the summer months. During June 1968, for example, temperature accounted for approximately 19% of the total vertical range in density at station 818P (mid-Bay off the mouth of the Patuxent).

Diurnally, variations as high as 3°C have been observed.[24] The annual range of temperature

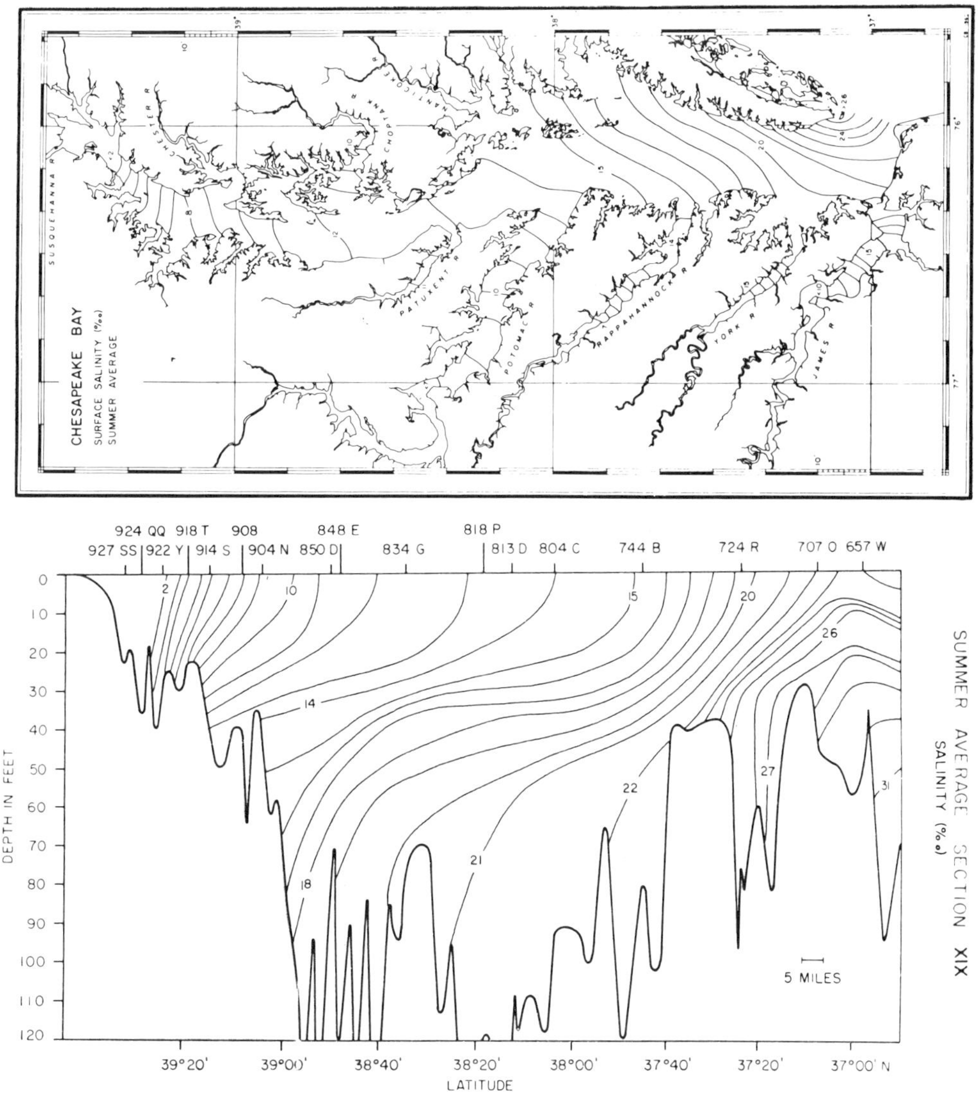

FIGURE 7. Horizontal distribution of average surface salinity in Chesapeake Bay — summer season.

in the open Bay is from about 1°C to approximately 29°C. There are also relatively large variations with periods in excess of 1 year. Daily measurements of surface temperature were taken for more than 50 years by the U.S. Coast and Geodetic Survey at selected tidal observation stations in some of the tributary estuaries.[25] Similar data are not available for the Bay proper, but comparison of the data, for say, Solomons, Maryland (Patuxent River) with Fort McHenry (Baltimore Harbor), suggests that these data are quite representative of the Bay system. An analysis of the Fort McHenry data set for the period 1914 to 1962 is given in Figure 9. The departure of the annual mean surface temperature at Fort McHenry from the long-term 49-year mean has been plotted on Figure 9. The figure shows that the annual mean had a range of 3.5°C, the maximum difference between consecutive years was greater than 1.5°C. It is also apparent from the figure that there are longer term oscillations present in the record.

Superimposed on these natural fluctuations are the thermal effects of man's activities, primarily the generation of electricity. The distribution of temperature is affected by man

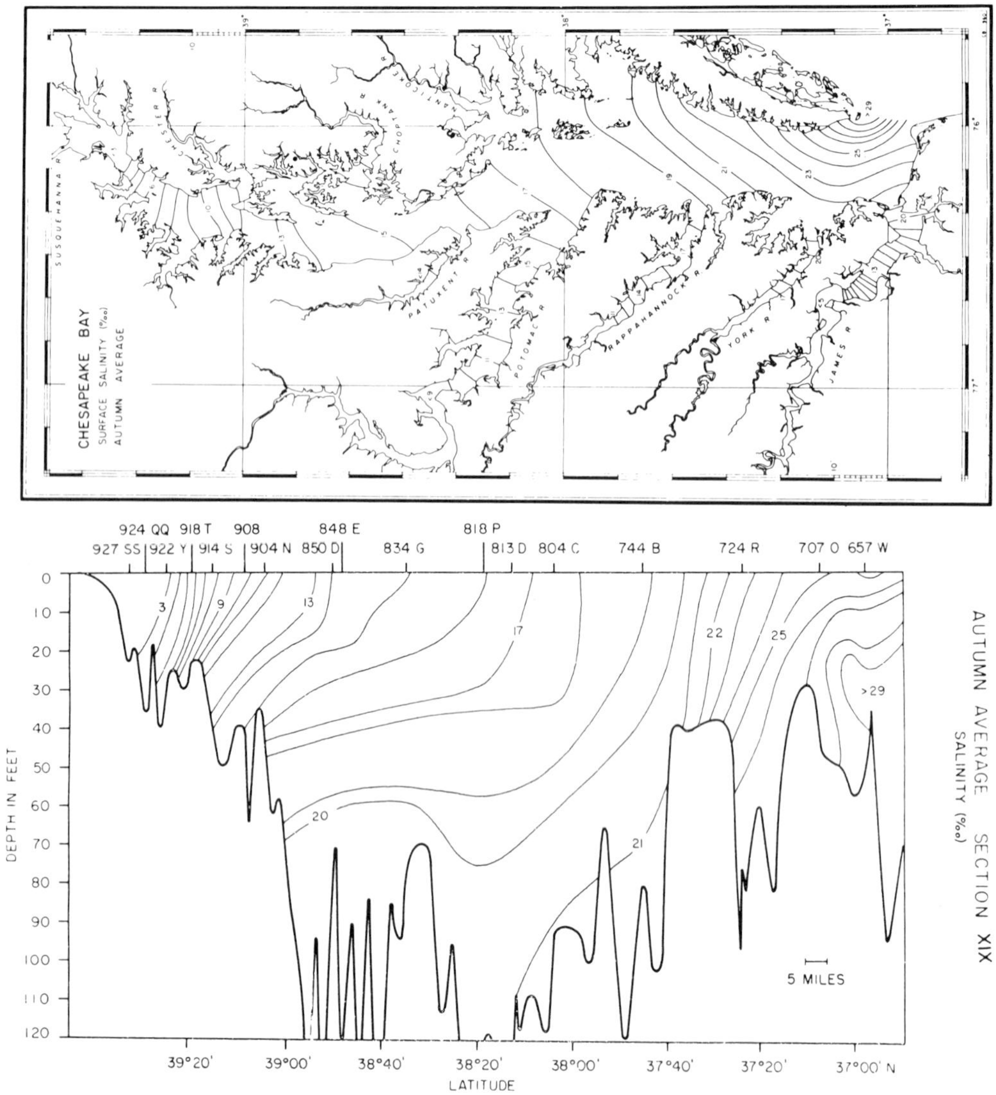

FIGURE 8. Horizontal distribution of average surface salinity in Chesapeake Bay — autumn season.

where large fractions of the available "dilution" water* is passed through the condensers of a generating station. Significant temperature effects due to this cause have not been demonstrated for the open Bay as yet but several tributary estuaries have had their temperature distributions measurably altered.

* Includes both freshwater runoff and water of oceanic origin.

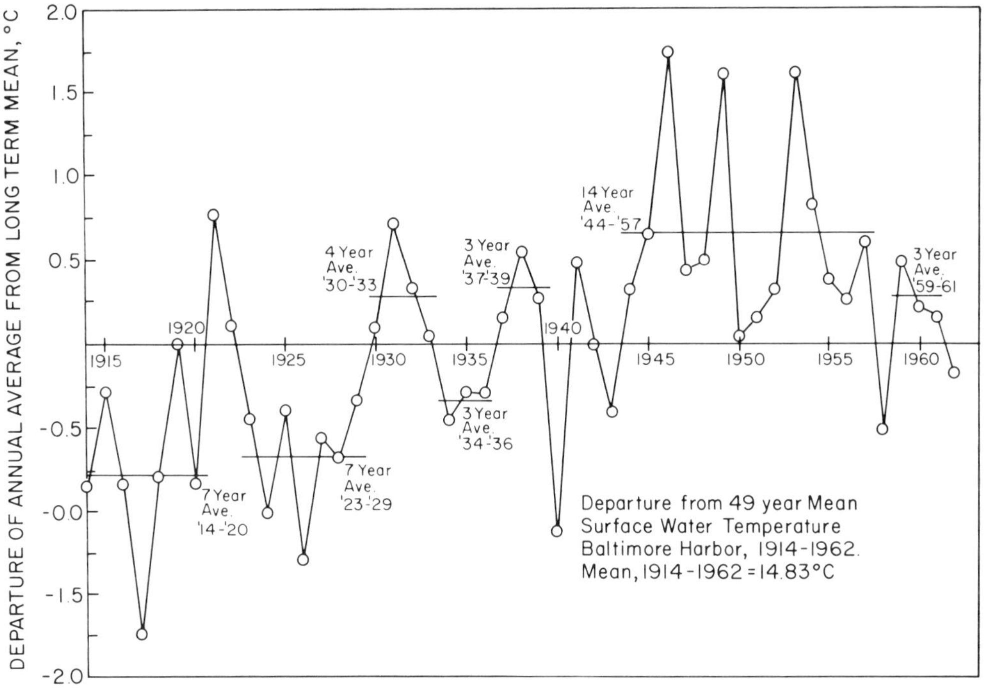

FIGURE 9. Departures of mean annual surface temperatures (°C) from 49-year (1914 to 1962) surface temperatures off Fort McHenry (Baltimore Harbor). Mean surface temperatures averaged over periods of several years are also shown.

REFERENCES

1. **Schureman, P.,** Tide and current glossary, Special Publication No. 228, U.S. Coast and Geodetic Survey, 1949, 7.
2. **Pritchard, D. W.,** Salinity distribution and circulation in the Chesapeake Bay estuarine system, *J. Mar. Res.*, 11(2), 106, 1952.
3. **Hicks, S. D.,** Tidal wave characteristics of Chesapeake Bay, *Chesapeake Sci.*, 3, 103, 1964.
4. **Elliott, A. J. and Wang, D-P.,** The effect of meterological forcing on the Chesapeake Bay: the coupling between an estuarine system and its adjacent coastal waters, in *Hydrodynamics of Estuaries and Fjords,* Nihoul, J. C. J., Ed., Elsevier, Amsterdam, 1978, 127.
5. **Elliott, A. J.,** A study of the effect of meteorological forcing on the circulation in the Potomac estuary, Special Report No. 56, Ref. 76-9, Chesapeake Bay Institute, The Johns Hopkins University, Bethesda, Md., 1976.
6. **Elliott, A. J.,** Observations of the meterologically induced circulation in the Potomac estuary, *Estuarine Coastal Mar. Sci.*, 6, 285, 1978.
7. **Elliott, A. J., Wang, D-P., and Pritchard, D. W.,** The circulation near the head of Chesapeake Bay, *J. Mar. Res.*, 4, 643, 1978.
8. **Elliott, A. J. and Hendrix, T. E.,** Intensive observations of the circulation in the Potomac estuary, Special Report No. 55, Ref. 76-8, Chesapeake Bay Institute, The Johns Hopkins University, Bethesda, Md., 1976.
9. **Wang, D-P. and Elliott, A. J.,** Non-tidal variability in the Chesapeake Bay and Potomac River: evidence for non-local forcing, *J. Phys. Oceanogr.*, 2, 225, 1978.
10. **Pritchard, D. W. and Rives, S. R.,** Physical hydrography and dispersion in a segment of the Chesapeake Bay adjacent to the Calvert Cliffs Nuclear Power Plant, Special Report No. 74, Chesapeake Bay Institute, The Johns Hopkins University, Bethesda, Md., 1979.
11. **Vieira, M. E. C.,** A Study of the Non-tidal Circulation in a Segment in the Middle Reaches of the Chesapeake Bay, Ph.D. Thesis, The Johns Hopkins University, Bethesda, Md., 1983.

12. **Pritchard, D. W. and Vieira, M. E. C.,** Vertical variations in residual current response to meterological forcing in the mid-Chesapeake Bay, in *The Estuary as a Filter,* Kennedy, V. S., Ed., Academic Press, Orlando, Fla., 1984, 27.
13. **Vieira, M. E. C.,** Estimates of subtidal volume flux in mid Chesapeake Bay, *Estuarine Coastal Shelf Sci.,* 21, 411, 1985.
14. **Pritchard, D. W.,** Chemical and Physical Oceanography of the Bay, Proceedings of the Governor's Conference on Chesapeake Bay, State of Maryland, II, 1968, 49.
15. **Pritchard, D. W. and Carpenter, J. H.,** Measurements of turbulent diffusion in estuarine and inshore waters, *Bull. Int. Assoc. Sci. Hydrol.,* 20, 37, 1960.
16. **Owen, W.,** A study of the hydrography of the Patuxent River and its estuary, Tech. Report No. 53, Chesapeake Bay Institute, The Johns Hopkins University, Bethesda, Md., 1969.
17. **Haas, L. W.,** The effect of the spring-neap tidal cycle on the vertical salinity structure of the James, York and Rappahannock Rivers, Virginia, U.S.A., *Estuarine Coastal Mar. Sci.,* 5, 485, 1977.
18. **Hayward, D., Welch, C. S., and Haas, L. W.,** York River destratification: an estuary-subestuary interaction, *Science,* 216, 1413, 1982.
19. **Pritchard, D. W.,** A study of the salt balance in a coastal plain estuary, *J. Mar. Res.,* 13, 133, 1954.
20. **Boicourt, W. C.,** The detection and analysis of the lateral circulation in the Potomac River Estuary, Maryland Power Plant Siting Program, Report PPRP-66, 1983.
21. **Schubel, J. R.,** The physical and chemical conditions of the Chesapeake Bay, *J. Wash. Acad. Sci.,* 2, 56, 1972.
22. **Pritchard, D. W.,** Chesapeake and Delaware Canal affects environment, Am. Soc. Civil Eng. Nat. Water Res. Eng. Meeting, Phoenix, Arizona, 1971.
23. **Seitz, R. C.,** Temperature and salinity distributions in vertical sections along the longitudinal axis and across the entrance of the Chesapeake Bay (April 1968 to March 1969), Graphical Summary Rep. No. 5, Ref. 71-7, Chesapeake Bay Institute, The Johns Hopkins University, Bethesda, Md., 1971.
24. **Beaven, G. F.,** Temperature and salinity of surface waters at Solomons, Maryland, *Chesapeake Sci.,* 1, 2, 1960.
25. U.S. Coast and Geodetic Survey, Surface water temperature and salinity, Atlantic Coast, Coast Geodetic Surv. Publ. 31-1, 1965.

Chapter 11

PUGET SOUND: A FJORD SYSTEM HOMOGENIZED WITH WATER RECYCLED OVER SILLS BY TIDAL MIXING

Curtis C. Ebbesmeyer, Jack Q. Word, and Clifford A. Barnes

TABLE OF CONTENTS

ABSTRACT

Puget Sound, in Washington State, is a system of fjord basins connected by constrictions in which there are strong tidal currents. The turbulence of these tidal currents causes water and other substances to be vigorously recycled among the basins before escaping the system after periods lasting as long as several years. Implications of recycling for man's activities need to be explored.

There are substantial variations of physical and biological characteristics at intervals of days, seasons, and years. Some of the processes which control the variability at the shorter intervals have been investigated, but those governing the longer scales remain poorly known.

I. INTRODUCTION

Along the coasts of Alaska, western Canada, Chile, New Zealand, and Norway, many of the world's fjord waterways are organized as systems of interconnected basins.[1-6] Some individual basins have been examined, but the behavior of overall fjord systems has received little study. Observations in one of these systems, Puget Sound in the northwestern U.S., indicate that the characteristics within a particular basin are strongly influenced by inputs throughout the system. The process responsible for this tendency toward homogeneity is intense tidal mixing in areas of land abutments (termed constrictions) which causes the water flow to be turbulent, thereby mixing surface and deep waters. These mixed water masses are recycled throughout the system. We present a summary of selected aspects of Puget Sound including physical characteristics, stratification and circulation in the basins, tidal mixing in the constrictions, and variability of physical and biological characteristics.

II. PHYSICAL CHARACTERISTICS

The Strait of Juan de Fuca extends eastward from the Pacific Ocean approximately 100 km where it branches into two fjord systems, the Strait of Georgia on the north[7] and Puget Sound on the south (Figure 1).

In the Puget Sound system, tidal mixing in areas of constricted water flow powers the circulatory patterns within the basins; five major regions have been selected according to the locations of major basins and constrictions (Figure 2). The primary entrance for water from the Pacific Ocean and exit for freshwater from Puget Sound is through a long throat (Admiralty Inlet). South of Admiralty Inlet three long, narrow basins diverge (Main, Hood Canal, and Whidbey basins). The last region (Southern Basin) connects to the Main Basin through another constriction. Figures 3 and 4 diagram the locations and depths of the basins and constrictions. With one exception (between basins B3 & B6), all of the constrictions contain a sill (S1-S9).

III. STRATIFICATION AND CIRCULATION IN THE BASINS

The patterns of water properties and currents within the basins fall into two groups: those having well-mixed, rapidly circulating water masses; and those having stratified, slowly circulating water masses (Figure 5). These differing characteristics are a result primarily of the presence or absence of constrictions and the relative input of freshwater. Two basins having well-mixed waters are the Southern and Main basins which receive 31% of Puget Sound's total runoff; there are four major constrictions where the water masses undergo turbulent mixing (S2, S3, S4, S5; Figure 3; Table 1). Two basins having stratified water masses (Hood Canal and Whidbey) receive 69% of Puget Sound's total runoff, however the inputs are not substantially mixed until they pass from these basins into Admiralty Inlet.

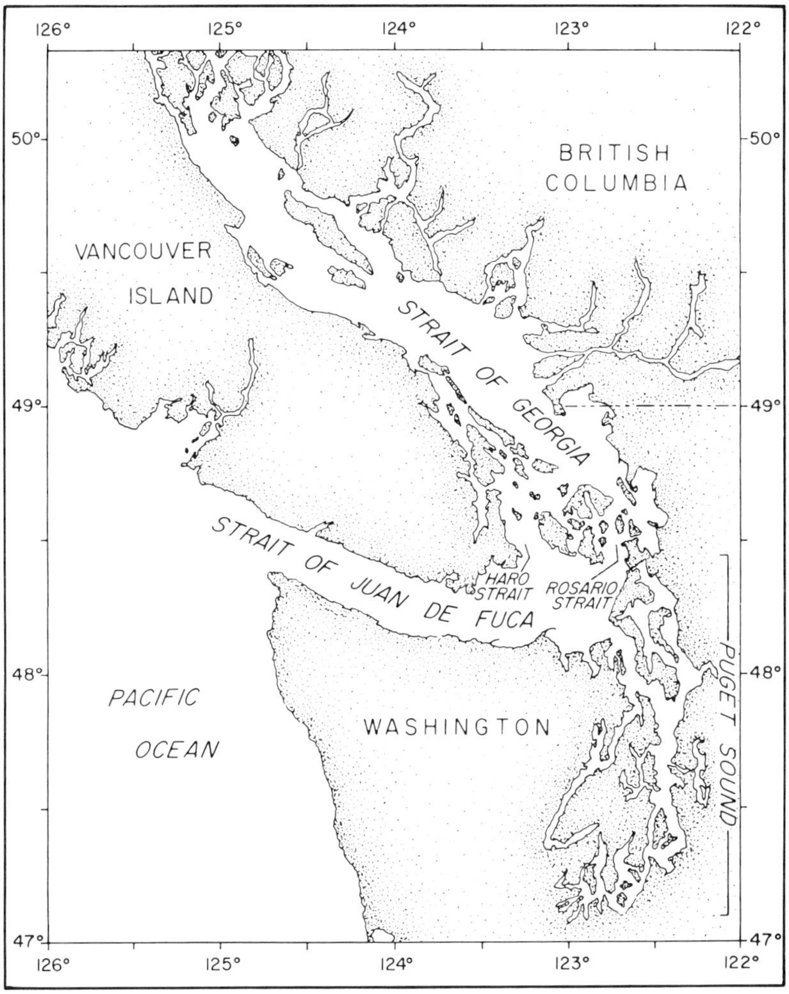

FIGURE 1. Inland marine waters of western Washington and Canada.

Volumetrically, the well-mixed basins account for 55% of Puget Sound's volume, whereas the stratified basins account for 32%; the remaining 13% is contained in Admiralty Inlet which is itself a constriction also characterized by vigorous mixing.

Because of mixing in the constrictions, waters in the Main and Southern basins have faster mean currents at depth than those in Whidbey and Hood Canal basins (Figure 5).[8] The speeds shown in Figure 5 are vector sums computed at particular depths over complete tidal cycles during intervals lasting days to months. These values represent the along channel current speed after fluctuations of tides have been suppressed by averaging. Net speeds are positive when oriented out of, and negative going into the fjord system. The principal outflow occurs in a layer in an upper part of the depth range and the inflow in a layer in the deeper depths. The horizontal pattern in these two flow layers is indicated throughout Puget Sound in Figure 6. There are a few exceptions where the flow is generally unidirectional from the water surface to the bottom as in the two channels in the southern half of the Main Basin (B3; Figure 3).

The outflowing layer occurs approximately in the upper 10 m for basins characterized by stratified water masses, whereas the outflowing layer extends much deeper (20 to 60 m) for those basins having well-mixed waters. The volume of water being transported in either

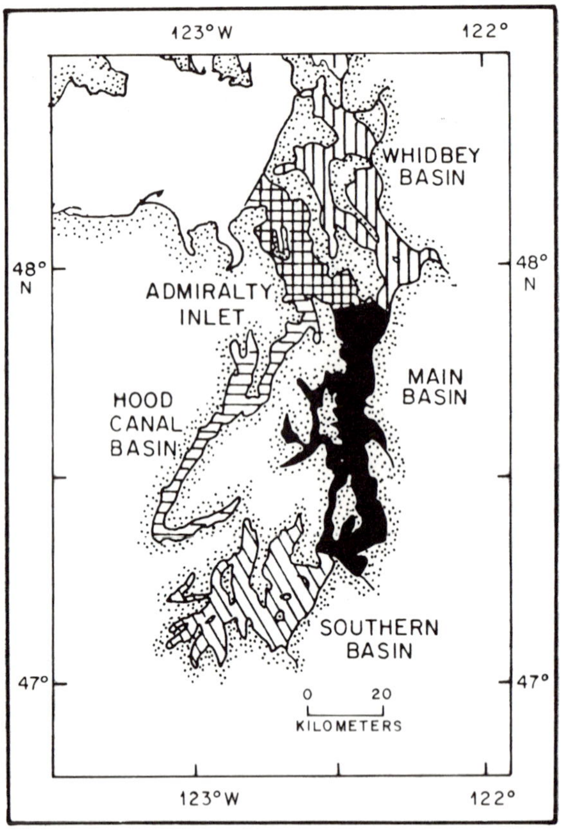

FIGURE 2. Five major subdivisions of Puget Sound. The entrance constriction (Admiralty Inlet; cross hatched) diverges into three basins (Hood Canal Basin, Whidbey Basin, Main Basin), and the Southern Basin connects to the Main Basin through a constriction. See Table 1 for physical characteristics of the divisions.

layer through a basin having well-mixed waters is an order of magnitude greater than for those having stratified water masses. For example, the volume transport of water flowing into and out of the Main Basin in the upper and lower layers is approximately 20,000 to 30,000 m^3/sec, and in the major channels of the Southern Basin the rate is 5,000 to 10,000 m^3/sec. In contrast, the volume transport is 1,000 to 2,000 m^3/sec for waters in Whidbey and Hood Canal basins.[9] These rates are much larger than the inputs from Puget Sound's rivers (1174 m^3/sec; Table 1), and enormous compared with inputs from man (order of 50 m^3/sec from all sewage treatment plants in Puget Sound).

IV. MIXING AND RECYCLING IN THE CONSTRICTIONS

Transport of water through the basins is primarily horizontal, whereas there is considerable vertical transport in the constrictions because of tidal turbulence. The relative intensity of the tidal mixing can be judged by comparing the kinetic energy of the currents throughout the system (Figure 4). The kinetic energy was determined by computing the total variance from numerous current meter records and averaging the results at selected locations. Kinetic energy in the constrictions is typically an order of magnitude larger than in the basins.

Because of the relative intensity of the tidal kinetic energy the majority of mixing in Puget Sound occurs in the constrictions, not the basins. Distinct water layers from the basins are

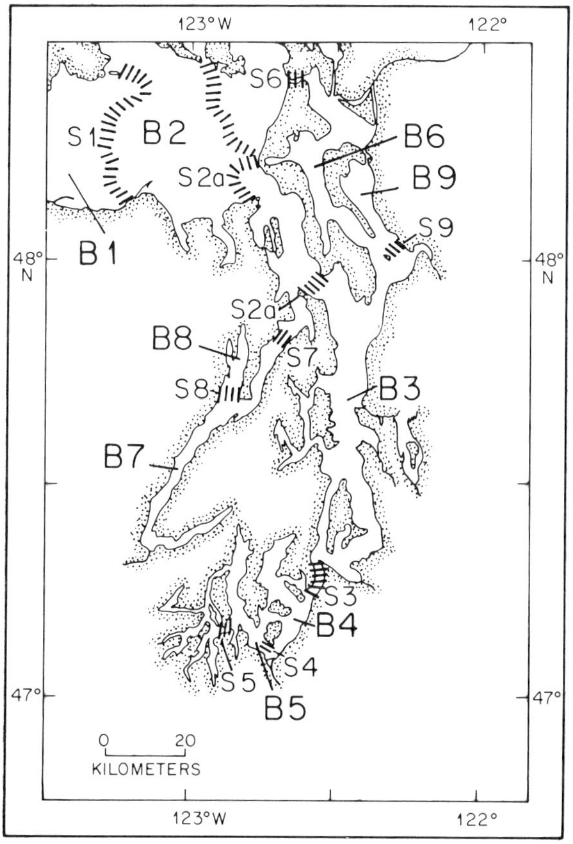

FIGURE 3. Horizontal view of Puget Sound showing sills (S1-S9; hatched) and basins (B1-B9). See Figure 4 for a vertical profile of midchannel bottom depths and locations of sills and basins.

joined at the constrictions by tidal mixing; it is here that water is primarily exchanged between layers. Water from a given layer at a constriction may travel in two directions (Figure 7): some is transported out of one basin and into another; and some is recycled in the constriction back into the other layer of the original basin. These two directions are illustrated in Figure 7 with lighter arrows indicating escape from a basin and heavier arrows indicating recirculation. The fraction of water recycled in a given constriction is typically on the order of one half;[10] thus, considerable recycling occurs in Puget Sound.

The process of recycling causes materials released in one basin to be exchanged throughout Puget Sound before escaping the system. The amount of time for escape to occur is a few months for approximately 50%, and 6 months for 90% of dissolved, inert substances.[11] Substances which rise or sink or are not inert have other behavior. The rising and sinking materials concentrate along edges of Puget Sound including its bottom, surface, and shorelines much like materials in bathtubs.[12]

V. PHYSICAL AND BIOLOGICAL VARIABILITY

The physical and biological characteristics of Puget Sound are dynamic and may fluctuate over time scales of days, months, and years. Sometimes variations may be extreme and yet overall effects are generally dispersed throughout the entire system because of the influence that recycling water masses have upon the passage of waters throughout this system. The

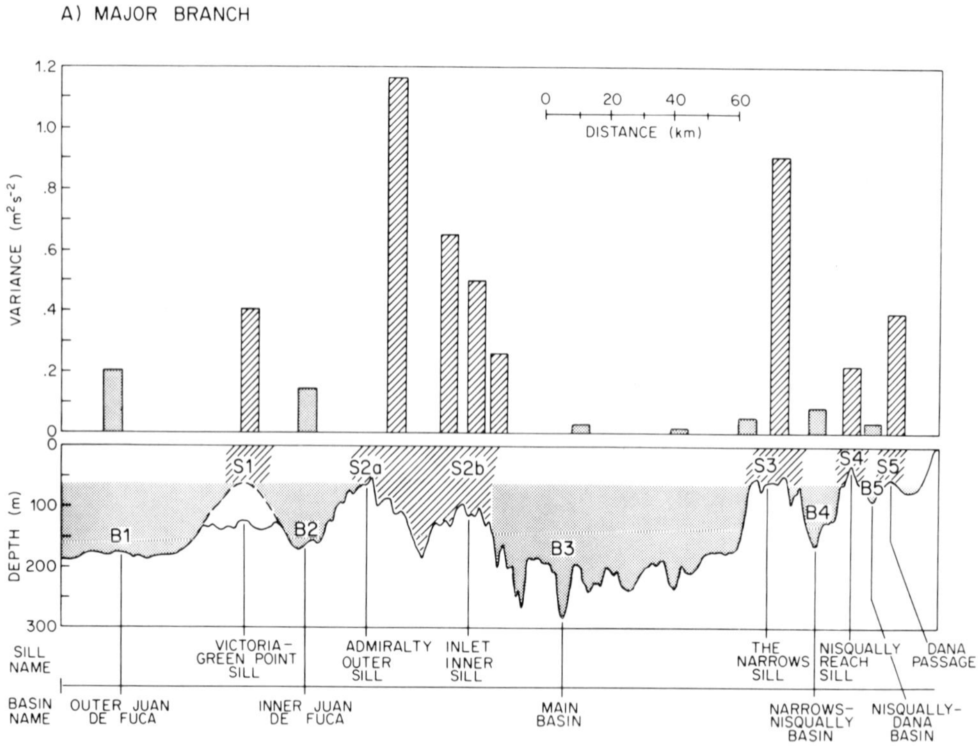

FIGURE 4. Profile view of midchannel bottom depth (lower panel) and total variance of currents at selected locations (upper panel) in Puget Sound: (A) Admiralty Inlet, Main Basin, Southern Basin; (B) Whidbey Basin; and (C) Hood Canal Basin. See Figure 3 for a horizontal view with designations for sills (S1-S9) and basins (B1-B9). For the origins of the place names see Phillips.[32]

variations considered are shown in Table 2. The values represent ranges: daily, the range of fluctuations over several days to several weeks; seasonal, the range between extremes of the seasonal cycles through a year; and interannual, the range between values seen in different years. These determinations were made for the Main Basin because this locale has been most intensively utilized by man. To compare the ranges of physical and biological characteristics, two dimensionless ratios were formed: the interannual range divided by the seasonal range, and the daily divided by the seasonal range. The seasonal range was used as the common denominator because it has been estimated for each of the selected characteristics.

The ratio of interannual to seasonal ranges varies between approximately 0.03 for benthic annelids to 2.0 for currents. The mean value is 0.6 based on 12 kinds of measurements. The ratio of daily to seasonal ranges varies from less than 0.2 for water temperature to more than 5.0 for runoff, currents, and chlorophyll. The mean value is 3.0 based on 10 samples. The observations indicate that physical and biological characteristics in Puget Sound fluctuate substantially on time scales of days to years.

Factors controlling daily and seasonal fluctuations have received some attention. Seasonal fluctuations of water temperature and salinity follow the seasonal cycles of air temperature and runoff, respectively.[11] Density driven currents near the bottom often occur at biweekly intervals and are related to tidal fluctuations at Admiralty Inlet.[13] Daily water temperatures at depth respond to fluctuations in local air temperatures.[14] Three fourths of the variance of daily average currents at 100-m depth can be explained by fluctuations of daily average winds,[15] and intense storms have dramatically altered water mass structure over a 200-m

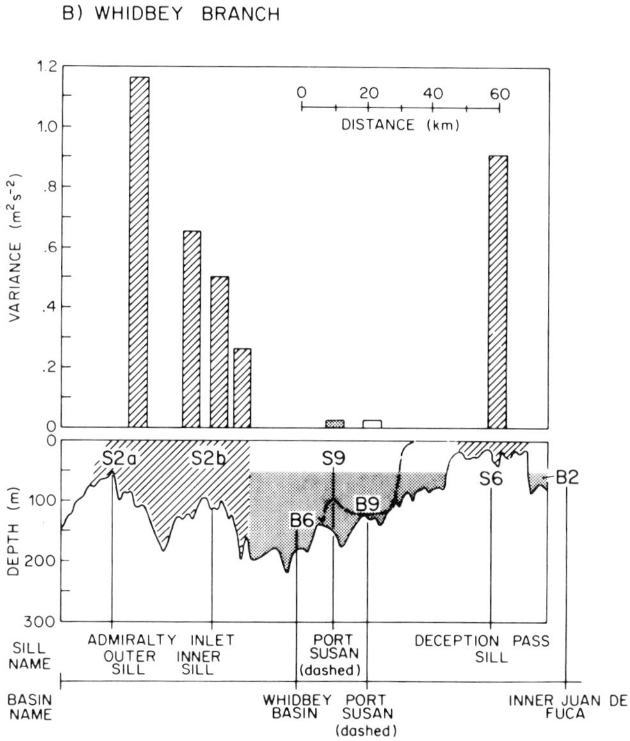
B) WHIDBEY BRANCH
VARIANCE (m²s⁻²)
1.2
1.0
.8
.6
.4
.2
0
0
20
40
60
DISTANCE (km)
DEPTH (m)
100
200
300
S2a
S2b
S9
B9
B6
S6
B2
SILL NAME
ADMIRALTY INLET
OUTER SILL
INNER SILL
PORT SUSAN (dashed)
DECEPTION PASS SILL
BASIN NAME
WHIDBEY BASIN
PORT SUSAN (dashed)
INNER JUAN DE FUCA

FIGURE 4B.

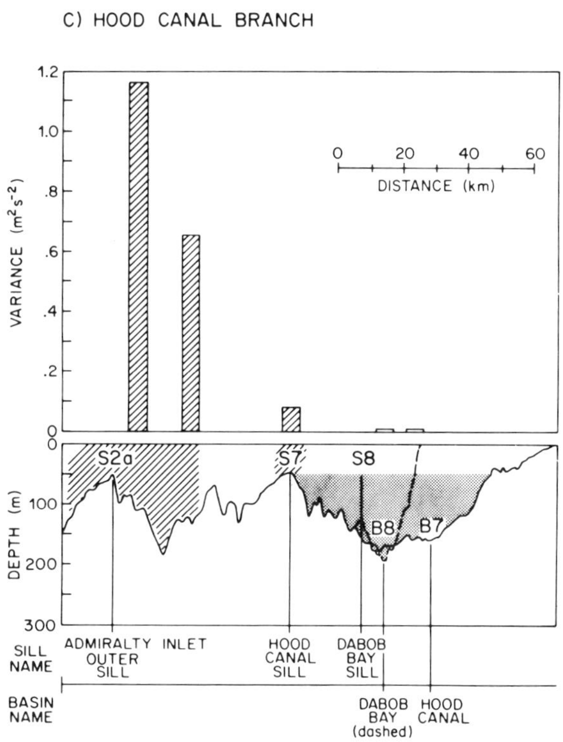
C) HOOD CANAL BRANCH
VARIANCE (m²s⁻²)
1.2
1.0
.8
.6
.4
.2
0
0
20
40
60
DISTANCE (km)
DEPTH (m)
100
200
300
S2a
S7
S8
B8
B7
SILL NAME
ADMIRALTY INLET
OUTER SILL
HOOD CANAL SILL
DABOB BAY SILL
BASIN NAME
DABOB BAY (dashed)
HOOD CANAL

C

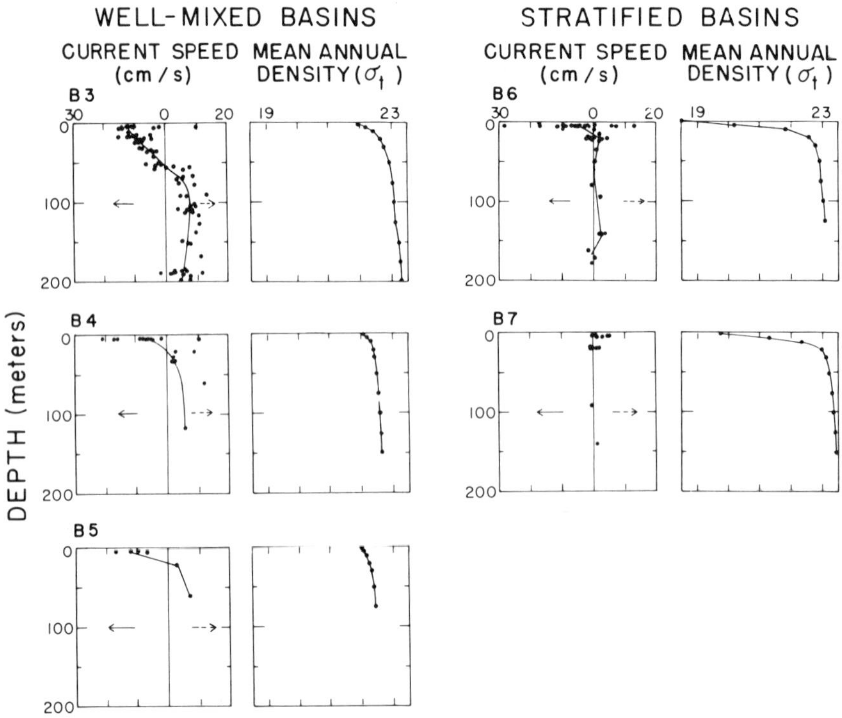

FIGURE 5. Density and current profiles in Puget Sound: at left, the well-mixed and rapidly circulating basins (Main and Southern basins); at right, the stratified and slower circulating basins (Hood Canal and Whidbey basins). Density profiles are annual averages of data obtained during 1932—1966 (see Collias[33]), and current data are vector average speeds from data of Cox et al.[34] Solid arrows indicate the direction pointing out of the fjord system, and dashed arrows the direction into the system (see Figure 6). Units: cm, centimeter; s, second, density expressed in σ_t units (grams per cubic centimeter − 1.000) × 1000.

depth range.[16,17] The daily fluctuations of chlorophyll and primary production can be explained using numerical models which incorporate local fluctuations in sunlight, tides, and nutrients.[18]

Some interannual fluctuations have been noted. Approximately three fourths of the interannual salinity and temperature variance can be explained by interannual variations of Puget Sound runoff and air temperature, respectively.[15] Between 1972 and 1981 the currents at 100-m depth decreased threefold in speed while the speed near the bottom showed a concomitant increase.[15] The abundance of molluscs at depths of 200 m have increased over the last 20 years.[19] Vertical profiles of sediment-bound metals correlate with man's activities over the past century.[20] Presumably some of these changes are linked to the recycling of man's inputs.

Puget Sound consists of 90% ocean water,[21] so that small fluctuations in oceanic conditions could also significantly alter conditions within Puget Sound. Further, the conditions in Puget Sound also depend on other inputs including those from the atmosphere and the Strait of Georgia. The blend of diverse inputs to Puget Sound's waters results in interannual fluctuations which remain poorly understood.

VI. CONCLUSIONS

The water within Puget Sound is a mixture of approximately 90% Pacific Ocean water and 10% freshwater which is vigorously cycled and recycled among the basins by tidal

Table 1
SELECTED PHYSICAL CHARACTERISTICS OF PUGET SOUND AND ITS MAJOR SUBDIVISIONS[f]

Characteristic	Entrance to Puget Sound Admiralty Inlet	Well-mixed basins: Main basin	Well-mixed basins: Southern basin	Stratified basins: Hood canal basin	Stratified basins: Whidbey basin	Entire Puget Sound
Shoreline length[a] (km)	178	533	620	343	469	2143
Area[a] (km^2)	395	767	448	388	634	2632
Volume[a] (km^3)	21.7	76.9	15.8	24.9	29.1	168
Mean tidal volume[b] (km^3)	0.997	2.43	1.68	1.14	1.82	8.07
Diurnal tidal[c] range (m)	2.56	3.48	4.39	3.60	3.51	2.56—4.39
Annual average[d] runoff (m^3/sec)	(e)	223	141	164	646	1174

[a] At mean high water.
[b] Between mean lower low water and mean high water.
[c] Diurnal range taken between mean higher high water and mean lower low water from summary by Mofjeld and Larsen[24] and National Ocean Survey.[25]
[a-c] Data from McLellan.[26]
[d] Average taken between 1930—1978 based on data from Coomes et al.[27] using the procedure of Lincoln.[28]
[e] Negligible runoff.
[f] See Figure 2.

mixing in the constrictions. This fjord system has a behavior which is distinct from that of single basins because of recirculating water masses.

The variability of Puget Sound's physical and biological characteristics is substantial on scales of days to months to years. Observations made continuously over a number of years throughout the system are needed to differentiate the processes governing variability at the various time scales.

The influence of man on Puget Sound has become evident as elevated metal concentrations in the sediments,[20] as tumors in fish in local embayments,[22] and as bacteria concentrations near the shore.[23] Because man's individual discharges are recycled throughout Puget Sound, the impact of these activities will have to be managed from a system point of view rather than according to individual discharges which is currently the practice.

ACKNOWLEDGMENTS

We thank Susan L. Ebbesmeyer for typing, Paul J. Ebbesmeyer and Donald M. Doyle for drafting, and Lucinda S. Word for editing.

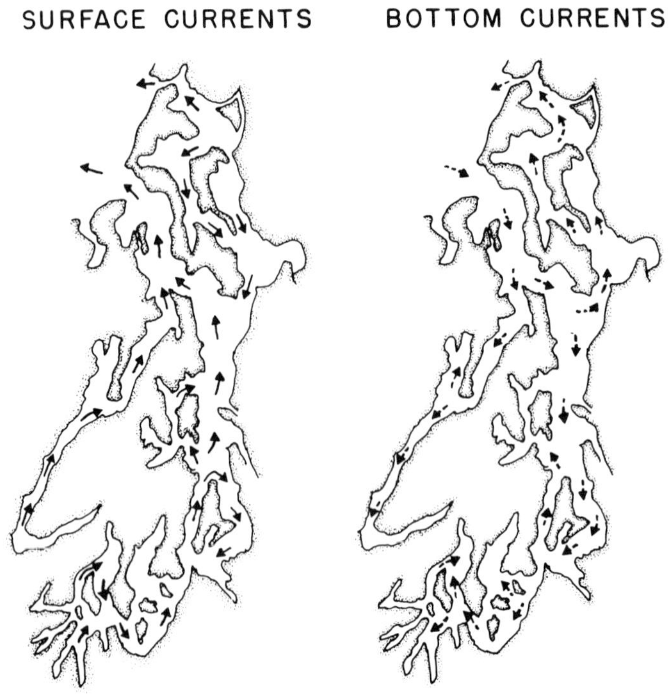

FIGURE 6. Mean flow pattern in Puget Sound: at left solid arrows indicate flow in the upper layer containing outflow, and at right dashed arrows indicate flow in the lower layer containing inflow. See Figure 5 for vertical profiles of current speeds.

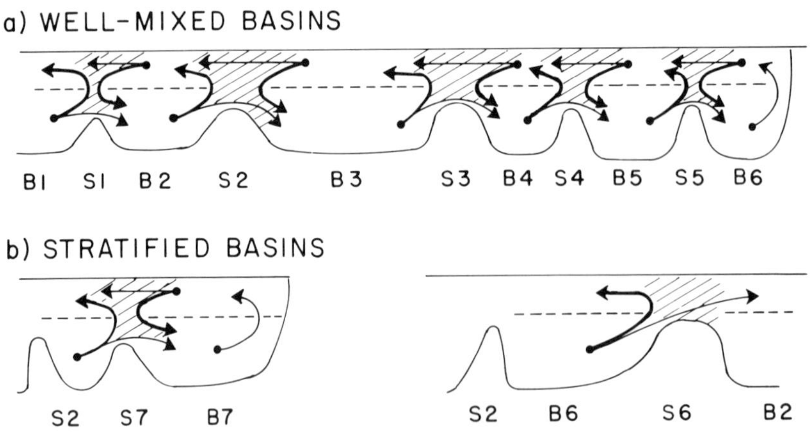

FIGURE 7. Schematic of refluxing in Puget Sound: a) well-mixed rapidly circulating basins; and b) stratified, slower basins. Darker arrows indicate recycling, or the amount mixed over a sill zone (hatched) and refluxed from one layer to another in a given basin. Lighter arrows denote the water which escapes from one basin to another through a constriction. See Figures 3 and 4 for sill (S1-S7) and basin (B1-B7) designations.

Table 2
VARIATIONS IN PUGET SOUND FOR SELECTED CHARACTERISTICS OVER TIME SCALES OF DAYS, SEASONS, AND YEARS

	Time scales			Ratios		
Characteristic	**Daily**	**Seasonal**	**Interannual**	**Inter-annual/ seasonal**	**Daily/ seasonal**	**Ref.**
Runoff (m^3/sec)	10,000	500—1800	700—1600	0.7	8	a
Air temperature (°C)	5	5—19	9.5—13.5	0.3	0.4	b
Wind speed (m/sec)	5	0.8—3.0	3	1	1	c
Water temperature (°C)						
5-m depth	1	8.1—13.1	9.8—11.6	0.4	0.2	d
100-m depth	1	7.9—11.5	8.9—10.7	0.5	0.3	d
Salinity (‰)						
5-m depth	2	28.3—30.2	28.5—29.7	0.6	1	d
100-m depth	0.4	29.5—30.6	29.7—30.3	0.5	0.4	d
Current (cm/sec)						
100-m depth	10— −30	6—9	4—10	2	13	e
Chlorophyll (mg/m^2)	10—200	10—40	?	?	6	f
Daily primary production (g Carbon/ m^2)	0.2—3	0.3—3	1—2	0.4	1	f
Benthic community at depth > 185 m (No. per 0.1 m^2)						
Molluscs	?	15—50	10—50	1	?	g
Arthropods	?	20—60	27—33	0.2	?	g
Annelids	?	30—60	38—39	0.03	?	g
Mean				0.6	3	
sample				12	10	

Note: Ratios are shown between the range of fluctuations at the time scales. Question marks indicate missing data, and references give data sources.

[a] Total runoff discharged into Puget Sound estimated by Coomes et al.[27] — daily, highest estimated during 1949—1978; seasonal, range of monthly averages computed for 1930—1978; interannual, range of annual averages computed for 1930—1978.

[b] Air temperature summarized by Coomes et al.[27] — daily, typical fluctuation of air temperature over several days recorded at Seattle-Tacoma International Airport; seasonal, range of monthly average air temperatures computed for Seattle during 1930—1978; interannual, range of annual average air temperatures computed for Seattle during 1900—1981.

[c] Wind summarized by Coomes et al.[27] — daily, standard deviation computed from average total variance of winds each day at Seattle-Tacoma International Airport (1949—1978); seasonal, range of average monthly net vector for 1949—1978 where positive is reckoned to the north and negative to the south; interannual, typical departure of a monthly vector net speed from the cycle of monthly averages.

[d] Water temperature and salinity near mid-channel off Point Jefferson and Shilshole Bay — daily, standard deviation of daily averages multiplied by four as computed for a current meter mooring during January 10—February 10, 1973[14]; seasonal, range of monthly averages;[11] interannual, range of annual averages computed for 14 1-year periods during 1938—1982.[15]

[e] Along channel current speed at 100-m depth at mid-channel off Shilshole Bay where positive is reckoned out estuary and negative, into estuary — daily, range of daily net speeds from records obtained during 1972—1977;[9] seasonal, range of monthly net vector;[15] interannual, range of trend estimated between 1972—1981.[15]

[f] Chlorophyll *a* and primary production integrated above the 1% light level off Point Jefferson and Shilshole Bay — daily, range of daily values observed during April—June, 1966 and April—May, 1967;[18] seasonal, range of monthly averages computed during 1966—1975 (see Strickland[29] after Ebbesmeyer and Helseth[30]); interannual, range of annual averages during 1966—1975.[30]

Table 2 (continued)
VARIATIONS IN PUGET SOUND FOR SELECTED CHARACTERISTICS OVER TIME SCALES OF DAYS, SEASONS, AND YEARS

[g] Total number of Mollusca at water depths in excess of 185 m between Brace and Browns points in East Passage. Seasonal range relates to samples taken in June, September, and November, 1982 and March 1983.[31] Interannual variation relates to comparison of the above information at a station south of Point Pully to data collected by Dr. F. H. Nichols (U.S. Geological Survey) in 1969—1970. Comparisons of equivalent sampling methods were made.[31]

REFERENCES

1. **Pickard, G. L.,** Oceanographic features of inlets in the British Columbia mainland coast, *J. Fish. Res. Board Can.*, 18, 907, 1961.
2. **Pickard, G. L.,** Oceanographic characteristics of inlets of Vancouver Island, British Columbia, *J. Fish. Res. Board Can.*, 20, 1109, 1963.
3. **Pickard, G. L.,** Some oceanographic characteristics of the larger inlets of southeast Alaska, *J. Fish. Res. Board Can.*, 24, 1475, 1967.
4. **Pickard, G. L.,** Some physical oceanographic features of inlets of Chile, *J. Fish. Res. Board Can.*, 28, 1077, 1971.
5. **Pickard, G. L.,** Annual and longer term variations of deepwater properties in the coastal waters of southern British Columbia, *J. Fish. Res. Board Can.*, 32, 1561, 1975.
6. **Nihoul, J. C., Ed.,** *Hydrodynamics of Estuaries and Fjords*, Elsevier, Amsterdam, 1978, 546.
7. **Waldichuk, M.,** Physical oceanography of the Strait of Georgia, British Columbia, *J. Fish. Res. Board of Can.*, 14, 321, 1957.
8. **Barnes, C. A. and Ebbesmeyer, C. C.,** Some aspects of Puget Sound's circulation and water properties, in *Estuarine Transport Processes*, Kjerfve, B., Ed., University of South Carolina Press, Columbia, S.C., 1978, 209.
9. **Ebbesmeyer, C. C., Coomes, C. A., Cox, J. M., Helseth, J. M., Hinchey, L. R., Cannon, G. A., and Barnes, C. A.,** Synthesis of current measurements in Puget Sound, Washington — Volume 3: Circulation in Puget Sound: an interpretation based on historical records of currents, *National Oceanic and Atmospheric Administration Technical Memorandum*, NOS OMS 5, U.S. Department of Commerce, Rockville, Maryland, 1984, 73.
10. **Cokelet, E. D., Stewart, R. J., and Ebbesmeyer, C. C.,** Exchange of water in fjords: application of the efflux/reflux theory to Puget Sound, *J. Geophys. Res.*, 1986, in preparation.
11. **Ebbesmeyer, C. C. and Barnes, C. A.,** Control of a fjord basin's dynamics by tidal mixing in embracing sill zones, *Estuarine Coastal Mar. Sci.*, 11, 311, 1980.
12. **Word, J. Q., Ebbesmeyer, C. C., Boatman, C. D., Finger, R. E., and Stober, Q. J.,** Vertical transport of effluent material to the surface of marine waters, *Oceanic Processes in Marine Pollution*, 1984, in press.
13. **Geyer, W. R. and Cannon, G. A.,** Sill processes related to deep water renewal in a fjord, *J. Geophys. Res.*, 87, 7985, 1982.
14. **Cannon, G. A. and Ebbesmeyer, C. C.,** Some observations of winter replacement of bottom water in Puget Sound, in *Estuarine Transport Processes*, Kjerfve, B., Ed., University of South Carolina Press, Columbia, S.C., 1978, 229.
15. **Ebbesmeyer, C. C., Cox, J. M., Coomes, C. A., Cannon, G. A., Stewart, R. J., Bretschneider, D., Holbrook, J., and Barnes, C. A.,** Interannual fluctuations of the velocity and mass structure of a well-mixed fjord, *J. Geophys. Res.*, in preparation.
16. **Kollmeyer, R. C.,** Water Properties and Circulation in Dabob Bay Autumn 1962, M.Sci. thesis, University of Washington, Seattle, 1965.
17. **Reed, R. J.,** Destructive winds caused by an orographically induced mesoscale cyclone, *Bull. Am. Meterol. Soc.*, 61, 1346, 1980.
18. **Winter, D. F., Banse, K., and Anderson, G. C.,** The dynamics of phytoplankton blooms in Puget Sound, a fjord in the northwestern United States, *Mar. Biol.*, 29, 139, 1975.
19. **Nichols, F. H.,** Abundance fluctuations among benthic invertebrates in two Pacific estuaries, *Estuaries*, 8, 136, 1985.
20. **Crecelius, E. A. and Bloom, N.,** Temporal trends of contamination of Puget Sound, *in* Oceanic Processes in Marine Pollution, 1984, in press.

21. **Friebertshauser, M. A. and Duxbury, A. C.,** A water budget study of Puget Sound and its subregions, *Limnol. Oceanogr.*, 17, 237, 1972.
22. **Malins, D. C., McCain, B. B., Brown, D. W., Sparks, A. K., and Hodgins, H. O.,** Chemical contaminants and biological abnormalities in central and southern Puget Sound, *NOAA Technical Memorandum,* OMPA-2, Boulder, Colorado, 1980, 285.
23. **Tomlinson, R. and Patten, M.,** Puget Sound monitoring program annual report, Water Quality Division, Municipality of Metropolitan Seattle, 1983.
24. **Mofjeld, H. O. and Larsen, L. H.,** Tides and tidal currents of the inland waters of western Washington, *National Oceanic and Atmospheric Technical Memorandum,* ERL PMEL-56, U.S. Department of Commerce, Seattle, Wash., 1984, 52.
25. National Ocean Survey, *Tide Tables 1984,* U.S. Department of Commerce, Boulder, Colo., 1983.
26. **McLellan, P. M.,** An area and volume study of Puget Sound, *University of Washington Department of Oceanography Technical Report,* 21, 39, 1954.
27. **Coomes, C. A., Ebbesmeyer, C. C., Cox, J. M., Helseth, J. M., Hinchey, L. R., Cannon, G. A., and Barnes, C. A.,** Synthesis of current measurements in Puget Sound, Washington — Volume 2: indices of mass and energy inputs into Puget Sound: runoff, air temperature, wind, and sea level, *National Oceanic and Atmospheric Administration Technical Memorandum* NOS OMS 4, U.S. Department of Commerce, Rockville, Md., 1984, 45.
28. **Lincoln, J. H.,** Derivation of freshwater inflow into Puget Sound, *Department of Oceanography Special Report,* 72, 20, 1977.
29. **Strickland, R. M.,** *The Fertile Fjord, Plankton in Puget Sound.* University of Washington Press, Seattle, 1983, 145.
30. **Ebbesmeyer, C. C. and Helseth, J. M.,** An analysis of primary production observed during 1966—1975 in central Puget Sound, Washington. Final Report to the Municipality of Metropolitan Seattle, 1976, 68.
31. **Word, J. Q., Striplin, P. L., Keeley, K., Ward, J., Sassano, J., Hulsman, S., Li, K., Sparks, P., Schroeder, J., LaValle, T., and Chew, K. K.,** Subtidal biology in Renton Sewage Treatment Plant Project: Seahurst Baseline Study, Submitted to Municipality of Metropolitan Seattle, 1983, 280.
32. **Phillips, J. W.,** *Washington State Place Names,* University of Washington Press, Seattle, 1982, 167.
33. **Collias, E. E.,** Index to physical and chemical oceanographic data of Puget Sound and its approaches, 1932—1966, *University of Washington Department of Oceanography Special Report,* 43, 823, 1970.
34. **Cox, J. M., Ebbesmeyer, C. C., Coomes, C. A., Helseth, J. M., Hinchey, L. R., Cannon, G. A., and Barnes, C. A.,** Synthesis of current measurements in Puget Sound, Washington. Volume I: Index to current measurements made in Puget Sound from 1908—1980, with daily and record averages for selected measurements. *National Oceanic and Atmospheric Administration Technical Memorandum* NOS OMS 3, U.S. Department of Commerce, Rockville, Maryland, 1984, 38.

Chapter 12

THE LAGUNA MADRE OF TEXAS: HYDROGRAPHY OF A HYPERSALINE LAGOON

Ned P. Smith

TABLE OF CONTENTS

I. INTRODUCTION

The Laguna Madre of Texas, together with the Laguna Madre Tamaulipas in northeastern Mexico, extend nearly continuously for 430 km along the northwestern rim of the Gulf of Mexico. The lagoon is characteristically a few kilometers wide and at most a few meters deep. The Texas portion of Laguna Madre is subdivided into northern and southern halves, separated by approximately 40 km of sand and mud flats which are only rarely inundated. The dredging of the Gulf Intracoastal Waterway in 1949 reconnected these two sections. At present, boat and barge traffic moves freely along the length of Laguna Madre, but the extent to which water is exchanged between the northern and southern portions of the lagoon is poorly understood.

Geological descriptions of Laguna Madre have been provided by Shepard and Moore,[1] Fisk,[2] and Behrens;[3] biological and ecological studies of the lagoon have been reported by Simmons,[4] Hildebrand,[5] Pulich,[6] and the Texas Department of Water Resources.[7] This paper emphasizes hydrography. Examples are restricted largely to the northern portion of the Laguna Madre of Texas, because this is the only segment of the lagoon for which adequate information exists to characterize hydrography as a response to the physical and climatological settings. The purpose of the paper is to summarize results of studies which provide insight into the effects of physical processes acting upon or within the lagoon. Integration of available information reveals a hypersaline lagoon, reflecting the semiarid climate, but modified by low frequency meteorologically forced exchanges with adjacent coastal bays.

II. CLIMATOLOGICAL SETTING

The regional climate of the coastal zone of south Texas is listed as tropical semiarid[8] and is anomalous enough to be included among the ''problem climates''.[9] There is some variability in reported values of annual precipitation and evaporation, perhaps indicating differences in the stations or time periods used in averaging. Climatic maps prepared by the Texas Water Development Board[10] indicate an average precipitation rate of approximately 75 cm/year (1931 to 1960), and a nearly identical average evaporation rate for the time period 1940 to 1957. Other climatological studies[11] indicate a north-to-south increase from 48 to 71 cm in the excess of annual potential evapotranspiration over direct precipitation. There is little freshwater inflow into northern Laguna Madre; inflow into the lagoon through Baffin Bay averages 1.0 m^3/sec and may cease altogether during periods of little or no rainfall.[12] Direct precipitation contributes an average of 65% of the total freshwater inflow to the lagoon.[7]

Several studies of the Texas coastal zone have looked specifically at departures from mean conditions. Behrens[13] distinguished between drought conditions (25 to 46 cm of rainfall per year), dry (46 to 61 cm), normal (61 to 79 cm), and wet years (more than 79 cm of rainfall). More recently, Norwine et al.[14] investigated trends and cyclical variations in precipitation superimposed onto the mean values, and showed quasiperiodic variations of 3 to 4 years in the annual precipitation totals. The meager annual precipitation and resulting freshwater runoff, coupled with an excess of evaporation, results in hypersaline environment which distinguishes Laguna Madre from many other coastal lagoons and from most estuaries in general. In negative estuaries, as in brackish water estuaries exhibiting the usual transition from fresh to marine conditions, salinity is a useful natural tracer for the movement of water.

III. HYDROGRAPHIC CHARACTERISTICS

Before the completion of the Gulf Intracoastal Waterway, Gunter[15] described the northern section of the Laguna Madre of Texas (Figure 1) as ''in good condition'' when salinities

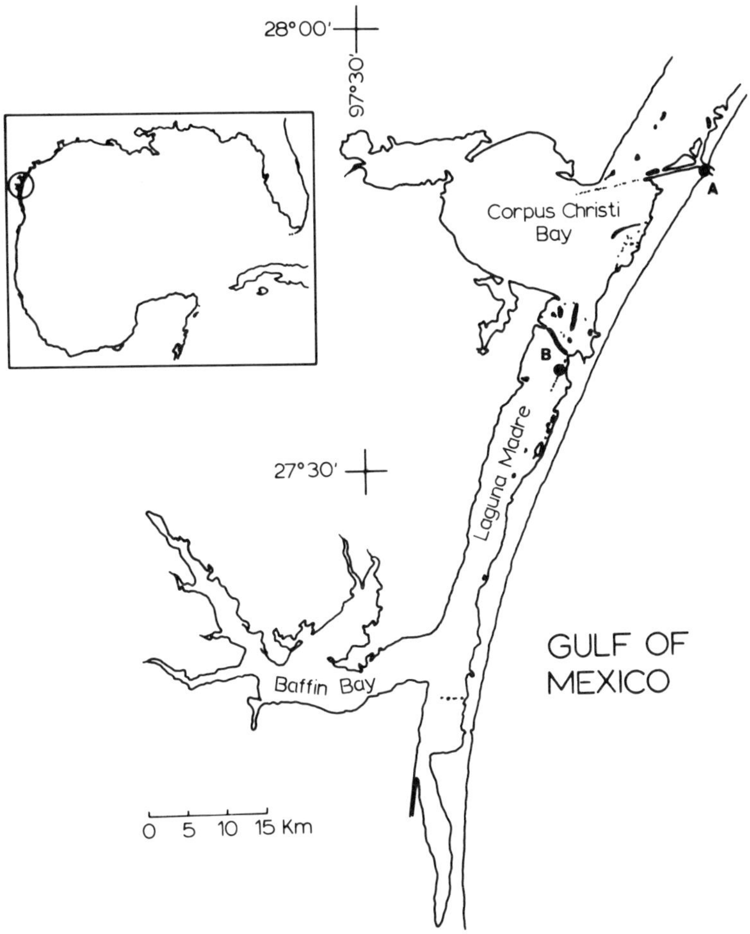

FIGURE 1. Map of the Laguna Madre, Corpus Christi Bay estuarine system. Insert shows the study area in the northwestern Gulf of Mexico. Stations A and B are shown at Aransas Pass and at the northern end of Laguna Madre, respectively.

fell within the range of 40 to 60 parts per thousand (ppt). During periods of unusually low rainfall, salinities approached 100 ppt. The Intracoastal Waterway improved the exchange of water with Corpus Christi Bay to the north, but Laguna Madre remains hypersaline, on average, at locations not normally reached by tidal and low-frequency exchanges.

Behrens[13] has described an extensive set of surface salinity data, collected along an approximately 50-km length of the Intracoastal Waterway, starting at the southern end of Corpus Christi Bay. In an earlier paper,[16] this data base was used to group salinities according to station, and to compute the mean and standard deviation for each station. Both the mean and standard deviation increased nearly linearly with increasing distance into the lagoon. The multi-annual mean salinity at the northern end of Laguna Madre was 31.5 ppt; this value increased southward at a rate of 0.18 ppt/km. The standard deviation increased at a rate of 0.10 ppt/km, starting at a value of 6 ppt at the northern end of the lagoon.

Temporal variations in salinity at selected sites along the Intracoastal Waterway are lost when available data are grouped to form averages, and the standard deviation is a poor indicator of temporal variability. Figure 2 is a mesh perspective which is somewhat qualitative in nature, but which reveals both temporal and spatial variations simultaneously during a period of time just over 2 years in length. The perspective selected has time, in days,

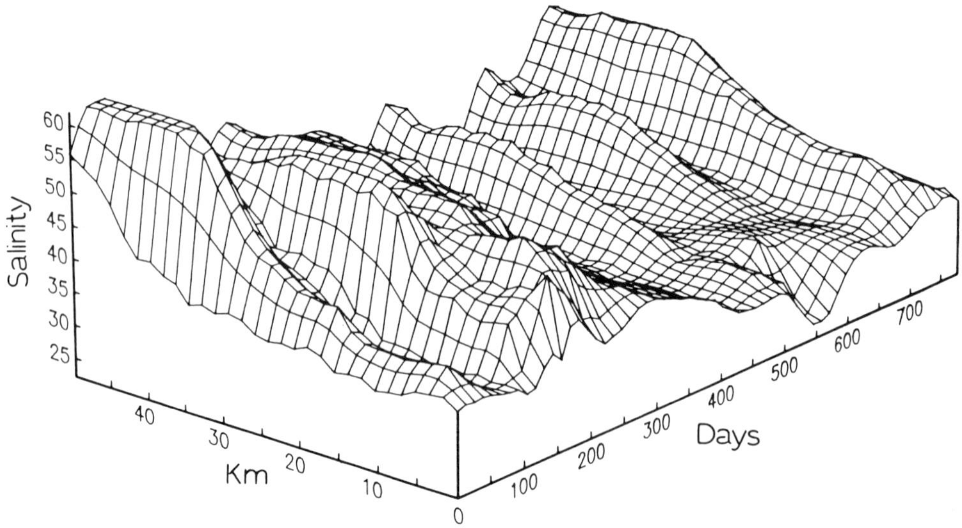

FIGURE 2. Mesh perspective plot of surface salinities through northern Laguna Madre, December 1964 to January 1967. Data were collected from 18 locations spaced approximately equally, starting at the southern extreme of Corpus Christi Bay.

increasing from left to right; distance into the lagoon, in kilometers, increases from front to back. It is clear that the buffering effect of exchanges with Corpus Christi Bay reduces temporal variations in salinity along the front of the plot, while the more extreme values are found in the interior of the lagoon along the back. A quasiperiodic variation in salinity, especially in the interior, indicates maximum values occurring approximately every 4 to 5 months.

IV. TIDAL AND LOW FREQUENCY WATER LEVEL VARIATIONS

Like most intracoastal lagoons with limited access to adjacent shelf waters, the Laguna Madre of Texas has characteristically low amplitude tides.[16] This is accentuated by the fact that in the northwestern Gulf of Mexico even coastal tidal amplitudes are relatively small.[17] Table 1 lists the harmonic constants of the principal tidal constituents computed from a 1-year water level record at the northern end of Laguna Madre, just south of Corpus Christi Bay. Amplitudes less than 1 cm are at or below the precision of the tide gauge and are therefore not included in the listing.

The relatively shallow water and narrow channels separating Laguna Madre from the Gulf of Mexico at Aransas Pass (Figure 1) act as an exponential filter, preferentially damping the higher frequency exchanges and water level variations. This is seen by comparing energy density and coherence spectra computed from water level records collected simultaneously at the coast and within the lagoon. Figure 3 contains energy density spectra computed from 148-day water level records collected at Station A at Aransas Pass (light line, with prominent spectral peaks) and at Station B at the northern end of Laguna Madre (bold line). Energy density levels at the longest periodicities are similar in magnitude, but the curves diverge with decreasing period. This is especially true at periodicities where spectral peaks appear in the Aransas Pass data. At the diurnal period, reflecting O_1 and K_1 tidal amplitudes, energy density levels computed from the Laguna Madre data are one and a half orders of magnitude lower than those computed from coastal water level data. Both of the principal diurnal constituents have amplitudes of approximately 11 cm at Aransas Pass.[18] Energy density levels associated with the M_2 constituent are two and a half orders of magnitude lower in

Table 1
HARMONIC CONSTANTS OF THE PRINCIPAL TIDAL CONSTITUENTS FOR THE NORTHERN SECTION OF LAGUNA MADRE, 8.2 KM SOUTH OF THE ENTRANCE TO CORPUS CHRISTI BAY

Constituent	Amplitude (cm)	Greenwich epoch (G)
Ssa	19.7	130
Sa	8.6	160
Mf	2.2	001
Mm	1.4	354
O_1	1.9	082
K_1	1.7	187

Note: Only constituents with amplitudes greater than 1 cm are listed. Calculations used a one-year water level record from February 1, 1974, through January 31, 1975.

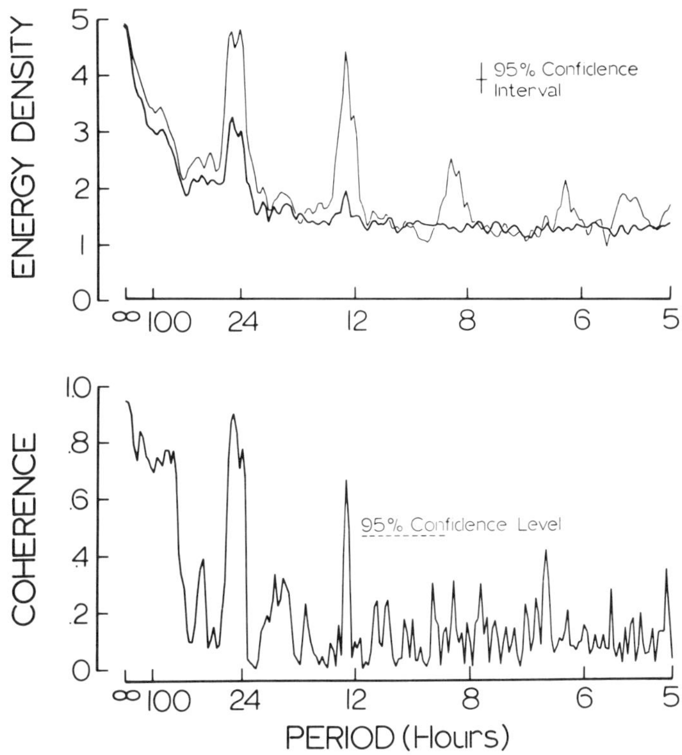

FIGURE 3. Composite of energy density spectra (upper) and coherence spectra (lower), computed from 149-day water level records collected simultaneously at Aransas Pass ("A" on Figure 1), and in northern Laguna Madre ("B" on Figure 1), January 30,—June 27, 1974. The energy density spectrum drafted with the darker line was computed using Laguna Madre data. Energy densities are in units of cm^2/c.p.h. Spectral resolution is 0.001 c.p.h.

Laguna Madre, and shallow-water overtides and compound tides are lost in the background noise completely. It is not known how tidal amplitudes decrease with increasing distance into northern Laguna Madre, but an analysis of a water-level record obtained 40 km to the south, at the entrance to Baffin Bay (see Figure 1), shows no indication of tidal period variations.

The coherence spectrum in the lower half of Figure 3 suggests a strong similarity between water-level variations in Laguna Madre and at the coast at both semidiurnal and diurnal tidal periodicities, and over time scales in excess of about 2 days. One may interpret this in terms of the rise and fall in water level at the coast forcing the tidal and long-period filling and draining of Corpus Christi Bay, and consequently of northern Laguna Madre. The coherence spectrum suggests that the exchange of water between Laguna Madre and the inner shelf occurs only at tidal periods and over time scales normally associated with meteorological forcing; the relatively small spectral peaks in the energy density spectrum computed from Laguna Madre data indicate that tidal exchanges play a distinctly secondary role in flushing the lagoon.

The tidal prism for Laguna Madre can only be estimated, because harmonic constants are available only from Station B (see Figure 1). Assuming a linear decrease in amplitude for the K_1 and O_1 constituents between Station B and the entrance to Baffin Bay (where amplitudes are known to be zero), the tidal prism varies from 5.78 million m^3 at times of tropic tides to 0.32 million m^3 at times of equatorial tides. Note, however, that these values may overestimate the true range considerably if tidal amplitudes are reduced to zero north of Baffin Bay. Also, the tidal prism indicates the total amount of water entering Laguna Madre between low tide and high tide; no correction is made for the source of the water, or for the net amount retained.

V. NUMERICAL MODELING OF BAY-SHELF AND INTERBAY EXCHANGES

As a result of the characteristically minor freshwater inflow into northern Laguna Madre, the flushing of the lagoon is primarily in response to exchanges with Corpus Christi Bay and, ultimately, with the Gulf of Mexico. Bay-Gulf exchanges, and exchanges between adjacent subestuaries, can be modeled adequately as barotropic processes, driven fundamentally by coastal water-level variations at Aransas Pass.[19] A simple, one-dimensional numerical model was developed to quantify the exchange of water between the inner shelf and Corpus Christi Bay, and then between the southern extreme of Corpus Christi Bay and northern Laguna Madre. Exchanges which reproduced observed water level variations were modeled as a balance between the barotropic pressure gradient and a quadratic frictional resistance term:

$$\Delta U/\Delta t = -\eta U\, g\, \Delta n/\Delta x - F\, u|U|$$

where U is the cross-sectionally averaged current speed in the channel connecting the two water bodies, $\Delta\eta$ is the difference in water level at two locations separated by a distance of Δx, g is gravity, and F is a frictional parameter with units of m^{-1} when current speeds are in m/sec. Simulated water-level variations were compared with analog water levels, and the friction term was found through trial-and-error to have a value of 1.43×10^{-4} for bay-shelf exchanges, and 1.22×10^{-4} for exchanges between Corpus Christi Bay and northern Laguna Madre. These constant values are clearly an over-simplification of a process which in reality will reflect bottom roughness, channel cross-sectional area and length; but these values were found to work satisfactorily. The correlation coefficient for calculated and measured volume changes of Corpus Christi Bay over a 49-day calibration period was +0.976.

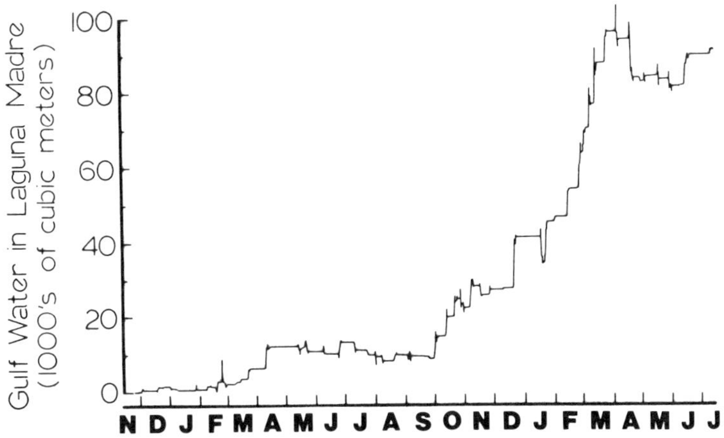

FIGURE 4. Time variation of accumulated Gulf of Mexico water calculated for northern Laguna Madre, November 14, 1973, through July 17, 1975. Values are in thousands of cubic meters.

Figure 4 shows the calculated net accumulation of Gulf water in northern Laguna Madre as a function of time during a 610-day simulation period from mid November 1973, through mid July 1975. Calculations assumed that Gulf water contained within the plume of water moving into the lagoon from Corpus Christi Bay mixed with lagoonal waters at a rate of 10%/hr. There is no indication of a tidal periodicity of any kind in the plot. Instead, variations in the fraction of Gulf water seem to occur as transient events — seemingly instantaneously on this compressed time scale. Because only Gulf water is considered in the calculations, the accumulation begins very slowly. During approximately the first 3 months, Gulf water is accumulating in Corpus Christi Bay, and little is transferred through the bay and into the lagoon. Also, this particular 3-month period (November to February) has characteristically falling water levels in the intracoastal bays and lagoons of South Texas.[18] Thus, the amount of Gulf water moving against a low frequency ebbing of water would be reduced. As Corpus Christi Bay fills with Gulf water, however, even modest bay-lagoon exchanges will advect greater amounts of Gulf water into the lagoon. As a result, later in the simulation, the rate of accumulation of Gulf water increases markedly. By the end of this particular 610-day period of time, a total of approximately 90,000 m^3 of Gulf water has been brought into Laguna Madre.

The net amount of Gulf water imported by tidal and low frequency exchanges can be put in perspective by comparing this volume with the total volume of the lagoon. Collier and Hedgpeth[20] have estimated the volume of Laguna Madre to be 174 million m^3 at mean tide level. The accumulation shown in Figure 4 represents 0.052% of this value — but it is important to keep in mind that the calculations were based upon assumed mixing rates for both Corpus Christi Bay and Laguna Madre. This study suggests that the renewal rate for Laguna Madre is slow, whatever the exact value might be. With such a slow long-term exchange, it is not surprising that the hydrography of the interior of the lagoon is so responsive to local precipitation-evaporation imbalances.

VI. DISCUSSION

An understanding of the hydrography of Laguna Madre requires results of investigations of air-sea interactive processes in a variety of forms, and over a wide range of time scales. The multi-annual excess of evaporation over direct precipitation and runoff explains the characteristic hypersalinity of the lagoon. Superimposed onto that, the interannual and seasonal variations in precipitation result in the quasiperiodic variations in salinity shown in Figure 2, as well as fluctuations too gradual to be resolved by the 26 months of available data. Over time scales ranging from days to weeks, variations in the windstress vector, occurring in response to synoptic-scale pressure patterns moving through the area, drive low-frequency exchanges which act to modify salinity extremes in the interior of the lagoon. Undoubtedly, shorter period variations would be identified if the sampling frequency were high enough. Smith[21] calculated substantial day-to-day variations in evaporation over a 68-day winter study carried out at the northern end of Laguna Madre.

The increase in both the mean salinity and the standard deviation southward along the axis of the lagoon is consistent with the mean salinity of Corpus Christi Bay, and these patterns can be explained in terms of the nature of exchanges between these two bodies of water. The bay has a substantial freshwater inflow[12] to counter the local excess of evaporation over direct precipitation. Holland et al.[22] have reported a mean bay salinity of 25.8 ppt computed from data collected monthly at 16 stations during the period from July 1973 through May 1975. The buffering effect of exchanges with the bay will extend into the lagoon as far as bay water is driven by tidal and meteorological forcing. Smith[16] used recording current meter data to document the displacement of bay water southward past a study site at the northern end of the lagoon. Tidal exchanges were negligible, but low frequency exchanges were significant. Unusually large displacements were correspondingly infrequent, but some displacements calculated during a 30-day study period in early 1976 suggested that bay water moved a substantial fraction of the distance from Corpus Christi Bay to Baffin Bay. More isolated locations in the interior of the lagoon would be influenced by lower bay salinities less often, and in the meantime the high evaporation rate would produce hypersaline conditions. Thus, the 40 ppt mean salinity found for the interior of the lagoon is no more surprising than the 10 ppt standard deviation. The seasonal and interannual alternation of wet and dry conditions documented by Behrens[13] drives the precipitation-evaporation imbalances which explain the temporal and spatial variations in salinity.

The relative importance of tidal flushing alone was examined by running the model with a time series constructed from the harmonic constants of the principal tidal constituents computed from the Aransas Pass record.[23] Results indicated that tidal forcing accounts for essentially none of the total. Apparently, even at times of tropic tides, tidal co-oscillations only move water back and forth within the channel connecting Laguna Madre with Corpus Christi Bay. To be effective as a flushing mechanism, tides must be superimposed onto longer period, nontidal processes.

Nontidal flushing forced by low frequency water level variations at Aransas Pass may account for only a fraction of the total nontidal flushing of the lagoon, however. The model does not take into account the low frequency set up and set down of water within the lagoon and within Corpus Christi Bay in response to local wind forcing. If a set up in water level at the northern end of Laguna Madre occurs simultaneously with a set down of water levels along the southern shore of Corpus Christi Bay, the resulting barotropic pressure gradient may produce a substantial exchange which could only enhance that which is driven by variations in coastal water levels. Such processes were not within the intended scope of the modeling exercise, but in view of the probability that they exist, the nontidal and total flushing quantified in this study should be thought of as minimum values.

Results of previous studies which have been integrated here suggest that, aside from its hypersalinity, the Laguna Madre of Texas is typical of many coastal lagoons. Tidal motions are greatly suppressed, due to the relatively small and widely spaced inlets, and due to the relatively large surface area of the lagoon itself. Only the longer-period processes, which will continue to fill or drain the lagoon for an extended period of time, can produce significant variations in water level, and thus effect an important flushing. The exchange of water between the lagoon and the adjacent continental shelf is directly proportional to both the magnitude of the forcing and the time scale over which it persists.

VII. SUMMARY

Salinity observations from a 26-month period of time in the mid 1960s are used to characterize the mean hypersaline state of northern Laguna Madre, Texas, and seasonal departures from the mean. Both the mean salinity and its standard deviation increase in the interior of the lagoon, away from the buffering effects of exchanges with adjacent, brackish water bays. At a location 50 km south of Corpus Christi Bay, the mean salinity is 40 ppt, and the standard deviation about the mean is 10 ppt. A three-dimensional mesh perspective plot suggests quasiperiodic variations in salinity over time scales of about 4 to 5 months occurring throughout the lagoon, but especially in the interior. Spectral analysis of lagoon and coastal water level records from early 1974 reveals statistically significant coherence levels at semidiurnal and diurnal tidal periodicities and over time scales in excess of about 2 days. Energy density spectra show that tidal period water level variations are greatly damped in the lagoon. Low-frequency meteorological forcing is of primary importance in driving lagoon-shelf exchanges, moderating salinity extremes and maintaining water quality. A 20-month coastal water level record from the mid 1970s is used in a simple, one-dimensional numerical model to quantify the transport of Gulf of Mexico water into the lagoon. Calculations suggest an intermittent renewal of water, primarily over fortnightly and seasonal time scales.

ACKNOWLEDGMENTS

I would like to express my appreciation to Mr. Mark Schmalz, who plotted Figure 2 and who wrote the computer model used to quantify total and tidal exchanges of lagoon and Gulf water. Water level records were provided by the Galveston, Texas, office of the U.S. Army Corps of Engineers. (Harbor Branch Foundation, Inc., Contribution Number 512.)

REFERENCES

1. **Shepard, F. and Moore, D.,** Central Texas coast sedimentation: characteristics of sedimentary environment, recent history and diagenesis, *Bull. Am. Assoc. Pet. Geol.*, 39, 1463, 1955.
2. **Fisk, H.,** Padre Island and the Laguna Madre flats, coastal south Texas, Second Coastal Geogr. Conf., Louisiana State Univ., Baton Rouge, La., 1959, 103.
3. **Behrens, E. W.,** Holocene sea level rise effect on the development of an estuarine carbonate depositional environment: Baffin Bay, Texas, *Mem. Inst. Geol. Bassin Aguitaine*, 7, 337, 1974.
4. **Simmons, E.,** An ecological survey of the upper Laguna Madre of Texas, *Contrib. Mar. Sci.*, 4(2), 156, 1957.
5. **Hildebrand, H.,** Laguna Madre Tamaulipas, observations on its hydrography and fisheries. Lagunas Costeras, un Symposio, Mex., D.F., Mem. Simp. Intern. Lagunas Costeras. UNAM-UNESCO, Nov. 28—30, 1967, 679.
6. **Pulich, W., Jr.,** Ecology of a hypersaline lagoon: the Laguna Madre, in Proceedings, Gulf of Mexico Coastal Ecosystems Workshop, Fore, P. and Peterson, R., Eds., U.S. Fish and Wildlife Serv. Publ. No. FWS/OBS-80/30, 1980.
7. Laguna Madre estuary: a study of the influence of freshwater inflows. LP-182, Texas, Dept. of Water Resources, Austin, Texas, 1983, 265.
8. **Critchfield, H.,** *General Climatology*, 3rd ed., Prentice Hall, Englewood Cliffs, N.J., 1974, 446.
9. **Trewartha, G.,** *The Earth's Problem Climates*, University of Wisconsin Press, Madison, Wis., 1961.
10. Monthly Reservoir Evaporation Rates for Texas, 1940—1957, Texas Water Development Board, Austin, Texas, 1960.

11. Physiography and climate map, Plate 1A, Texas General Land Office, Coastal Management Program, Austin, Texas, 1975.
12. **Diener, R.,** Cooperative Gulf of Mexico estuarine inventory and study — Texas: area description. NOAA Tech. Rept, NMFS CIRC-393, U.S. Department of Commerce, 1975, 129.
13. **Behrens, E. W.,** Surface salinities for Baffin and Laguna Madre, Texas, April 1964—March 1966, *Publ. Inst. Mar. Sci.*, 11, 168, 1966.
14. **Norwine, J., Bingham, R., and Zepeda, R.,** Twentieth-century semi-arid and subhumid climates of Texas and northeastern Mexico, Proceedings Int. Conf. on the Meterol. of Semi-Arid Zones, Tel Aviv, Israel, 1977, 30.
15. **Gunter, G.,** Some characteristics of ocean waters and Laguna Madre, *Texas Game and Fish*, 7, 19, 1945.
16. **Smith, N. P.,** Tidal and long-period exchanges between upper Laguna Madre and Corpus Christi Bay, Texas, Vol. 11, Texas A & I University Studies, 1978, 37.
17. **Zetler, B. and Hansen, D.,** Tides in the Gulf of Mexico — a review and a proposed program, *Bull. Mar. Sci.*, 20, 57, 1970.
18. **Smith, N. P.,** Intracoastal tides of upper Laguna Madre, Texas, *Tex. J. Sci.*, 30, 85, 1979.
19. **Smith, N. P.,** Numerical simulation of bay-shelf exchanges with a one-dimensional model, *Contrib. Mar. Sci.*, 28, 1, 1985.
20. **Collier, A. and Hedgpeth, J.,** An introduction to the hydrography of tidal waters of Texas, *Contrib. Mar. Sci.*, 1(2), 121, 1950.
21. **Smith, N. P.,** Energy balance in a shallow seagrass flat for winter conditions, *Limnol. Oceanogr.*, 26, 482, 1981.
22. **Holland, J., Maciolek, N., Kalke, R., Mullins, L., and Oppenheimer, C.,** A benthic and plankton study of the Corpus Christi, Copano and Aransas Bay systems, Third Year Report, Univ. Texas Mar. Science Inst., 1975, 174.
23. **Pore, N. and Cummings, R.,** A fortran program for the calculation of hourly values of astronomical tide and height of high and low water. U.S. Weather Bureau (now National Weather Service). Tech. Memorandum TDL-6, Silver Spring, Md., 1967.

Chapter 13

THE MOBILE BAY ESTUARY: STRATIFICATION, OXYGEN DEPLETION, AND JUBILEES

William W. Schroeder and Wm. J. Wiseman, Jr.,

TABLE OF CONTENTS

I. INTRODUCTION

Jubilees, mass migrations of estuarine organisms to the shores of Mobile Bay, have been presumed to be induced by hypoxic bottom waters. A review of new and historic data sets corroborates this hypothesis and elucidates the processes causing hypoxia. Hypoxia may be frequent and widespread during the summer, stratified season, while the winter climatology appears generally to preclude its occurrence.

Mobile Bay (Figure 1) is a semienclosed basin located on the northern coast of the Gulf of Mexico (30.5° N, 88°W). Geomorphologically, it is a combination of the drowned river valley and bar-built estuarine types. It has a surface area of 984 km^2 and an average depth of approximately 3 m. It drains a watershed of 11.3×10^4 km^2 and is the terminus of the Mobile River system, which has the sixth largest discharge on the North American continent.[1] The Mobile River system carries the combined flows of the Alabama and Tombigbee Rivers, which represent approximately 95% of the freshwater input into Mobile Bay. The average flow of the system is approximately 1800 m^3 sec^{-1}, and the 10 and 90 percentile flows are approximately 4250 and 370 m^3 sec^{-1}, respectively.[2,3] The lowest 7-day average flow on record is 223 m^3 sec^{-1}, which occurred in 1954.[4] The highest 7-day average flow on record is 13,977 m^3 sec^{-1}, which occurred in April 1979.[5]

Interactions of the Bay's geomorphology, water column structure, circulation, biological activity, and man-made modifications result in oxygen depletion zones. These, in turn, elicit a movement response from certain estuarine biota. The relationships between these factors are the subject of this paper.

II. GEOMORPHOLOGY

There are two openings in the southern part of the Bay: Main Pass, which provides a direct connection to the Gulf of Mexico, and Pass-aux-Herons, which opens onto the east end of Mississippi Sound. The exchange of waters with Mobile Bay through these two passes is estimated to be 85 and 15% of the total exchange, respectively.[2] The astronomical tides are principally diurnal, with a mean diurnal range of approximately 0.4 m, a maximum tropic tide range of 0.8 m, and a minimum equatorial tide range of 0.0 m.

Considering its shallow depths, Mobile Bay is bathymetrically complex (Figures 1 and 2). A dredged ship channel, 120 m × 12 m, cuts through the Bay from Main Pass to the Port of Mobile. Midway up the Bay, two shorter channels branch off to the west. The 5.2 km^2 triangular-shaped Gaillard Island, located between these two channels, was constructed in the early 1980s from dredged material. Shoal areas occurring along the western side of the main ship channel result from the historic practice of open-water disposal of dredged material.

Deep holes, with depths greater than 5 m, are found throughout the central area of the northeastern part of the Bay. The exact origin of these features is unknown. Surveys from the 1800s indicate that this area of the Bay may have been, naturally, slightly deeper than adjacent areas. However, the data are sketchy. The most likely contributor to the existence of these deep holes appears to be shell dredging that occurred in the Bay from 1946 to 1982 and which removed a reported 4.7×10^7 m^3 of clam and oyster shell.

III. STRATIFICATION

Since the work of Austin,[6] it has been known that strong haline stratification occurs within the waters of Mobile Bay. Stratification levels in excess of 5‰/m are reported by Austin[6] and also by Schroeder and Lysinger,[7] a level similar to that found in certain salt wedge estuaries.[8] This fact and its implications have not generally been acknowledged. Broad,

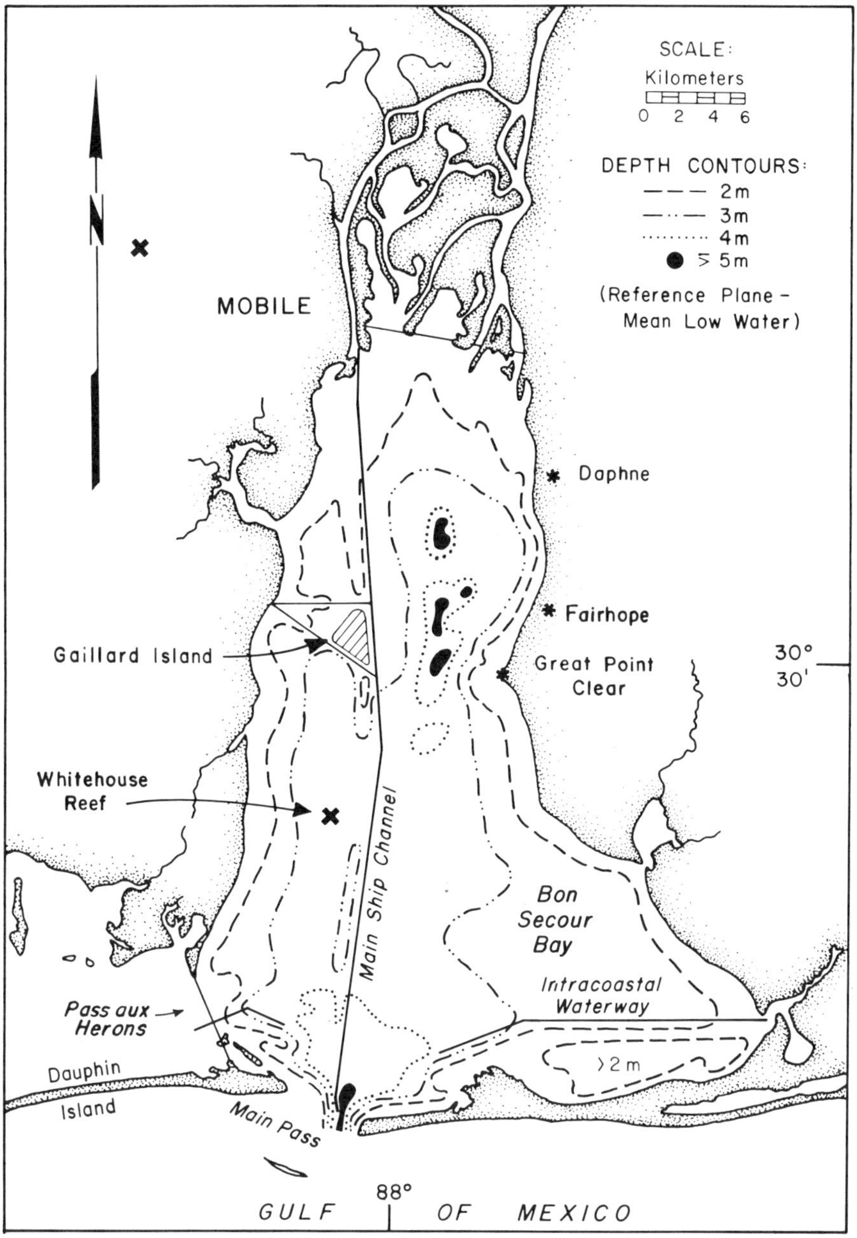

FIGURE 1. Bathymetric chart of Mobile Bay.

shallow estuaries are still often assumed *a priori* to be vertically well-mixed (e.g., Ling[9]). Attempts to model the Bay outside the ship channel have assumed vertical homogeneity (e.g., April et al.[10] and Youngblood and Raney[11]). The presence of haline stratification over the shallow parts of the Bay has profound effects upon the way the system functions, irrespective of whether the salt enters through a natural or dredged channel.

Strong thermal stratification has not been observed in Mobile Bay on a regular basis. Schroeder and Lysinger[7] report that the monthly averaged thermal vertical structure in the upper half of the Bay undergoes an annual reversal, with generally stable gradients during August through January, unstable gradients from February through June, and homogeneous

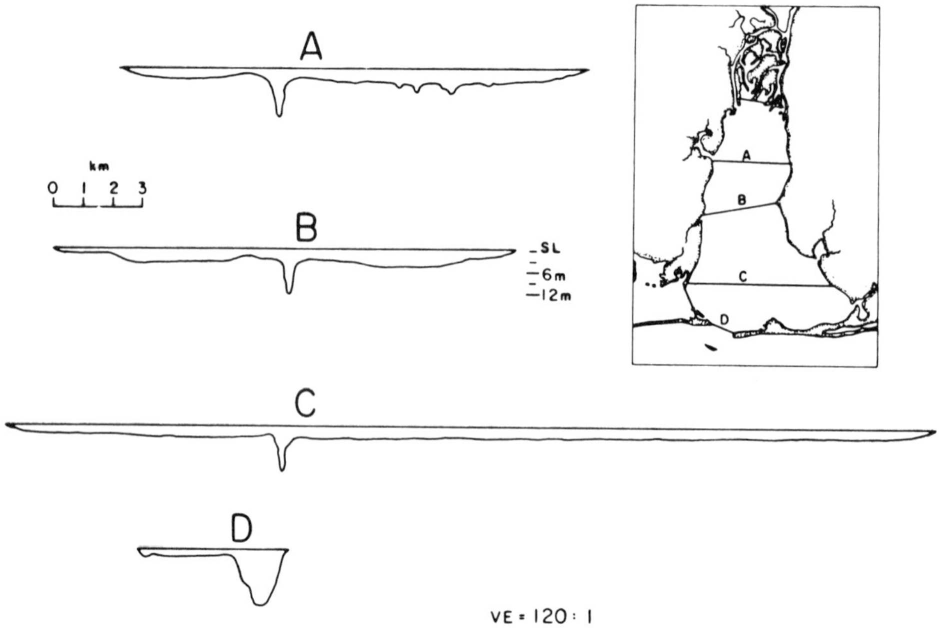

FIGURE 2. Selected bathymetric cross sections of Mobile Bay. Inset: location map.

conditions in July. During the periods of stratification the average differences between surface and bottom temperatures were equal to or less than 1.0° C except during April, when the bottom waters were 1.5° C warmer than the surface. Nevertheless, density stratification is always stable, and is controlled primarily by salinity.

Salinities within the Bay change on a number of different time scales. At tidal periodicities, the salinities of both the upper and the lower layers exhibit large variations. When stratification is strong, though, the motion in the two layers may contain significant independent components. Thus, both the absolute salinities and the surface-to-bottom salinity difference change during a tidal cycle (Figure 3) rather than just the former as one tends to find in a partially mixed estuary where barotropic tidal advection dominates.

Subtidal variability is also important. Salinity time series from the summer of 1978 (Figure 3) suggest periods of 5 to 20 days to be particularly energetic. With the short records available, though, it is difficult to identify the ultimate cause of these fluctuations. Wind stress, river runoff, and the fortnightly tides all vary on similar time scales.

Between July 3 and 10, 1978, the bottom salinities at Fairhope were recorded as less than the surface salinities. Thermal stratification was insufficient to ensure a stable water column during this event. The instrument was operating properly both at the beginning and at the end of the deployment, as evidenced by comparison with deck STD units. It is possible that local artesian ground water seeps affected the data, but without further evidence, the salinity inversion must remain unexplained.

Stratification is often intense along the length of the Bay, generally ranging between 5 and 10‰ over the depth of the shallows outside the ship channel. Figure 4 shows examples from early September 1980 of strong halocline structures along both a longitudinal salinity section next to the ship channel and a lateral section midway up the Bay. Runoff was very low (<500 $m^3 sec^{-1}$). Winds were less than 5 m sec^{-1} for the 3 days preceding the sections. This stratification is responsive to the wind. Following storms, the Bay becomes vertically isohaline, but the stratification recovers on time scales of the order of a few days to 1 week. On seasonal scales, the stratification responds to the riverine flood cycle.[3]

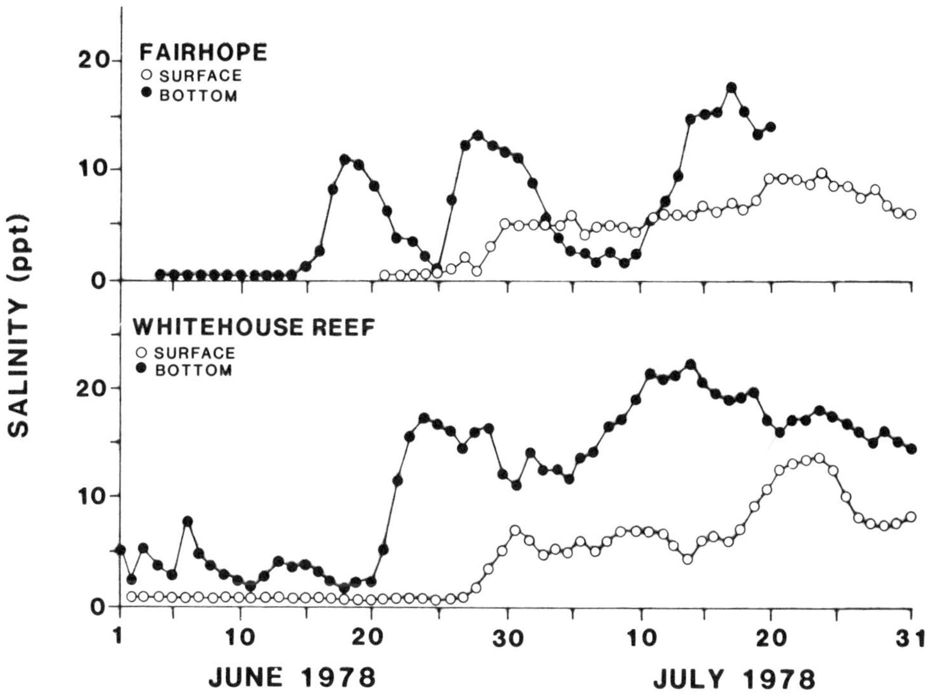

FIGURE 3. Time-series of daily mean salinities near Fairhope and Whitehouse Reef.[7]

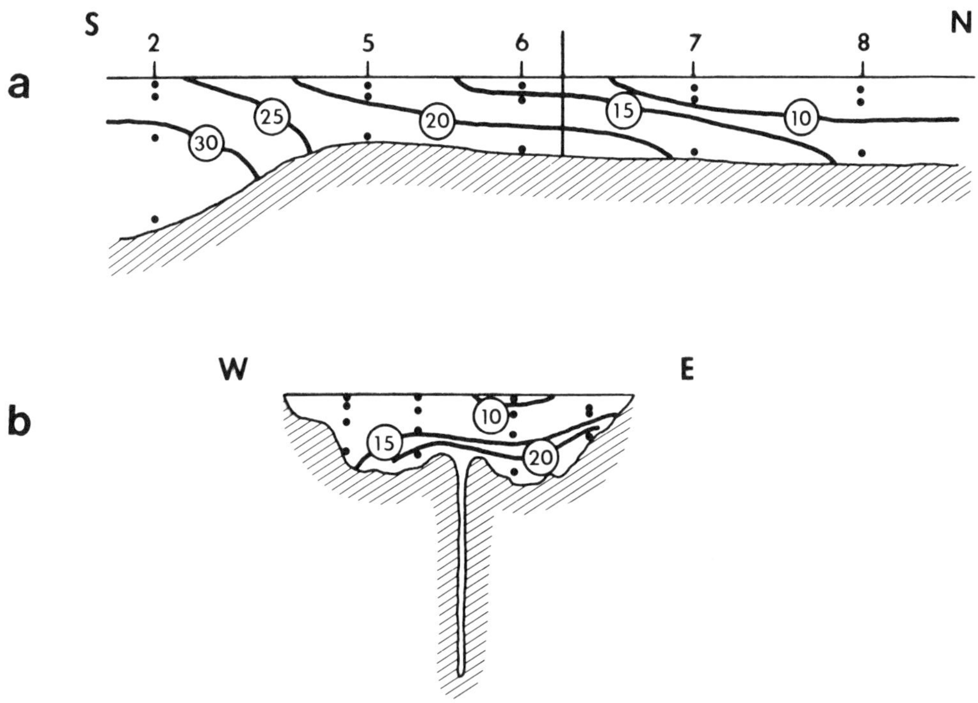

FIGURE 4. (a) Longitudinal salinity section immediately east of the ship channel, September 8, 1980.[19] Vertical line indicates position of cross section shown in b. (b) Lateral cross-section of salinity taken September 9, 1980.[20] Dots represent data points.

IV. OXYGEN DEPLETION AND "JUBILEES"

The aperiodic mass migrations of certain estuarine organisms to the upper eastern shore region of the Bay were linked by both Loesch[12] and May[13] to the occurrence of low oxygen concentrations in the deeper waters of the Bay. When these animals reach the shallow waters at the shoreline, they often remain there in a deteriorating or dying state for up to several hours but are seldom killed.[13] This phenomenon is locally called a "jubilee".

Jubilees, so named around 1912, are considered by most laymen to be grand social events whereby nature has provided for easy harvest of crab, flounder, and shrimp. They have been observed only during the summer months and nearly always take place in the early morning hours before sunrise. Newspaper accounts as far back as July 17, 1867 have recorded the occurrence of this phenomenon and indicated that it was an annual event.[13]

Periods of oxygen depletion in Mobile Bay were first reported in the scientific literature by Loesch,[12] with additional documentation by May[13] and Schroeder.[14] Loesch's[12] discussion is limited to three data sets collected during two different summer seasons in the vicinity of Daphne, Alabama (Figure 1), with only two to eight observations each. He reported dissolved oxygen values of less than 1.0 ppm on two occasions. May's[13] work covered the entire summer season, June through September, of 1971 and encompassed sampling locations throughout the bay. No specific data are presented, but rather a composite map of the minimum dissolved oxygen concentration observed at each sampling station. Oxygen levels less than 3.0 ppm affected an area of 445 km^2 or 44% of the Bay. May states that extensive oxygen depletion (values less than 1.0 ppm) occurred in July and August somewhere in the bay on 75% of the days sampled. This severe hypoxia was usually restricted to the bottom 1 m of the water column and was found with equal frequency on both the east and the west sides of the ship channel.

Schroeder[14] presents data from both Mobile Bay and east Mississippi Sound documenting oxygen depletion that accompanied an extreme river flooding episode during the spring (April 16 to May 15) of 1973. His results illustrate the progression of events associated with increasing haline stratification and decreasing oxygen concentrations. Although oxygen concentrations decreased throughout the entire water column in Mobile Bay, the lowest values (2.5 to 3.0 ppm) were observed only in the higher salinity bottom waters of the lower Bay near Main Pass (Figure 1). It is suggested that this is the result of increased vertical stratification brought about by the flooding river waters reaching the lower portion of the Bay and the accumulation of oxygen-consuming organic material near the sediment-water interface. Similar conditions of decreased oxygen concentrations were recorded during river flooding episodes in 1979[5] and 1980.[15] No "jubilee-type" activities have been reported during any of the hypoxic periods associated with these flooding events.

Both Loesch[12] and May[13] present plausible hypotheses as to the cause and extent of the summer season low oxygen conditions. Unfortunately, neither investigator was working with a very extensive data base, and therefore both were left in a mostly speculative position. Historically, the environmental conditions that have been associated with the occurrences of jubilees are listed below. Some of these factors are clearly contributory to the summer formation of hypoxic conditions.

1. Bottom water stagnation is caused by a stratified water column resulting from low-salinity upper Bay waters overlying higher salinity lower Bay waters, often with only a few parts per thousand differences. This isolates the bottom water from reoxygenation by gas exchange across the air-sea interface.
2. This stagnation occurs first, and to a greater degree, in areas with isolated deep bathymetric features or with restricted circulation as a result of shoaling. Both conditions may result from natural or anthropogenic activities.
3. The presence of large accumulations of rotting organic debris principally deposited in

the Bay during the previous spring river flood enhances oxygen consumption in the bottom water.

4. Light (5 m sec^{-1}) to calm winds over the days preceding and easterly winds on the day previous to and during the jubilee result in surface waters being moved away from the eastern shore. Lower-layer hypoxic water moves toward the eastern shore to maintain continuity. The jubilee ceases if the wind direction changes.
5. A rising tide just prior to and during the jubilee apparently results in bottom waters being forced toward the eastern shore beaches. If the tide reverses and starts to fall, the jubilee ends.

It appears reasonable to accept the linkage between periods of oxygen depletion and the occurrence of jubilees even though the level of knowledge concerning these two phenomena is rudimentary. Because of the obvious difficulties associated with attempting to predict either event, observation of their occurrence has tended to be serendipitous rather than planned.

Such was the case in July 1978 when one jubilee event was reported 6 days prior to a scheduled field sampling program and a second jubilee took place 4 to 8 hr before the first field survey. During the early morning hours of July 11, 1978, flounders, crabs, eels, and sting rays were reported caught in large numbers along the eastern shore north of Great Point Clear (Figure 1) (The Mobile Press Register, July 16, 1978). Then again on the morning of July 17, 1978, a jubilee was reported along the beaches near Daphne, Alabama (Figure 1).[16] Hydrographic and meteorological data collected in and around Mobile Bay are presented herein and form the basis for the most comprehensive documentation to date of Mobile Bay under conditions of severe oxygen depletion.

Figure 5 illustrates the bottom salinity and bottom dissolved oxygen fields obtained from a sequence of four surveys in the upper Bay over 40 hr during July 17 to 18, 1978. Each survey corresponded with the period immediately surrounding either predicted high or low water. From late afternoon on July 16 to the evening of the 17th the winds were generally to the south at 3 to 5 m sec^{-1}. They then diminished and became weak and variable until midday on the 18th, at which time they increased to 2 to 5 m sec^{-1} to the north and northeast. River input was generally low, with discharges less than 500 m^3 sec^{-1}.

The bottom salinity fields (Figure 5) resemble a north-south oriented tongue of intermediate salinity (8 to 16‰) water. As expected, the higher salinity values most frequently occur on the deeper, eastern side of the Bay. This tongue migrates up and down the Bay in response to flooding and ebbing tides, respectively. The modifying influence of the wind on the structure of the bottom salinity fields can be seen by contrasting Figure 5a and 5e at high tide periods and Figure 5c with 5g at low tide periods. In Figure 5a, north and northeasterly winds prevent the higher salinity bottom waters from moving into the upper Bay, as happens in Figure 5e, when the winds are light and variable. Also, in Figure 5a the bottom isohalines are shown compacted towards the east as a result of the winds pushing low salinity surface water to the western side of the Bay which results in eastward compensatory movement of the bottom waters. We believe that the lobe of 12 to 16‰ water, shown extending from the middle of the Bay up against the eastern shore (Figure 5a), illustrates the mechanism required to move oxygen deficient (<1.0 ppm) bottom waters (Figure 5b) up to the shoreline thus setting the stage for the reported jubilee earlier that same day.

During low tide and light and variable winds, the tongue is found withdrawn to the south (Figure 5c). In Figure 5g southerly winds have apparently held the higher salinity waters in the central region of the Bay during low tide and piled up surface waters in the upper end of the Bay, resulting in lower salinity bottom waters (<8‰) being moved to the south along the western and eastern shores.

The bottom dissolved oxygen fields (Figure 5) are visually coherent with the bottom

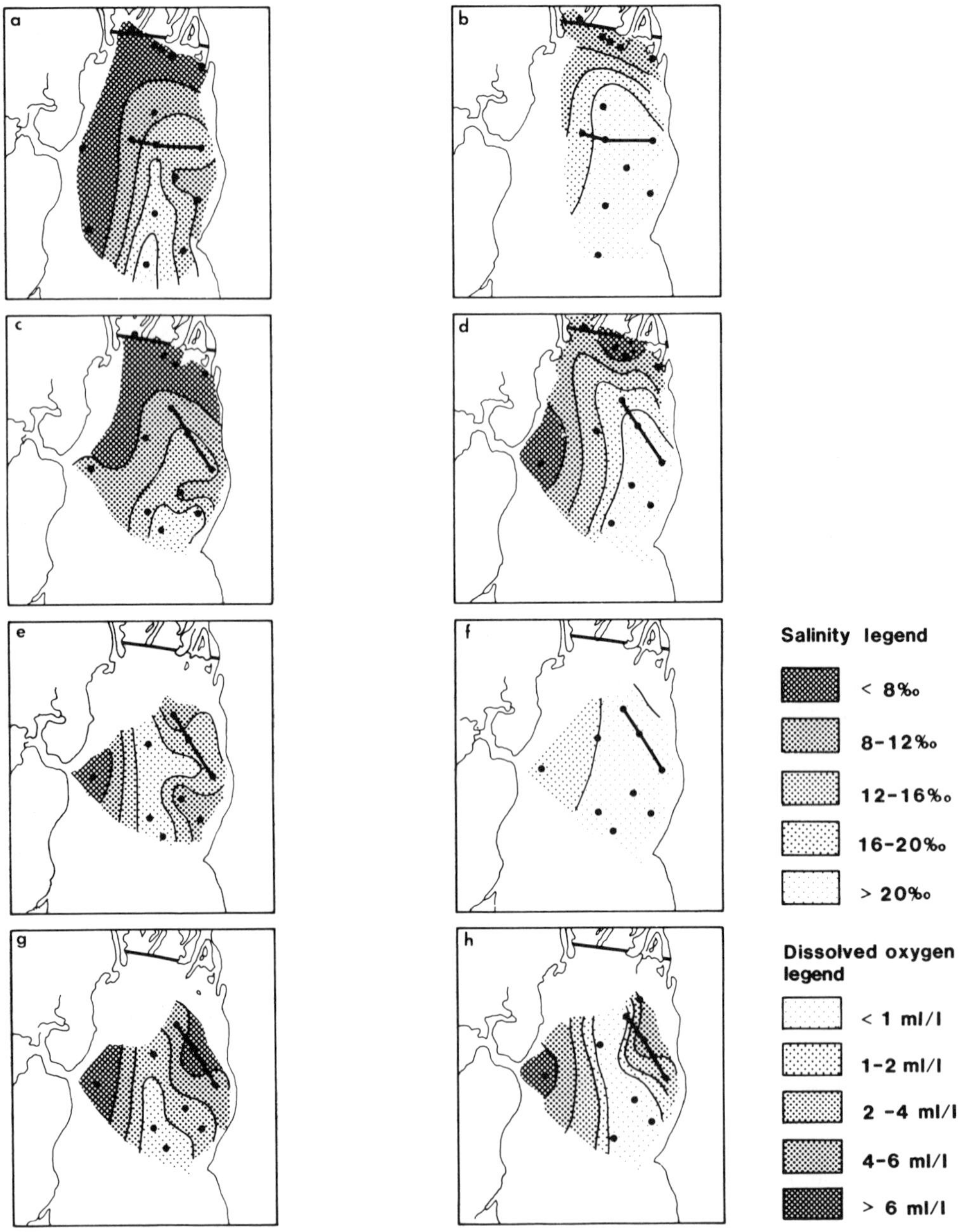

FIGURE 5. Bottom salinity (a, c, e, g) and corresponding dissolved oxygen (b, d, f, h) fields for consecutive high and low tide periods: 1099-1245 CDT July 17, 1978 (a, b), 2100-2355 CDT July 17, 1978 (c, d), 0945-1235 CDT July 18, 1978 (e, f), 2055-2359 CDT July 18, 1978 (g, h). The series begins with high water. Dots represent station locations. The straight lines connect stations used in constructing vertical sections of Figure 6.

salinity fields. Salinities greater than 12‰ were always associated with dissolved oxygen values less than 2 $m\ell\ \ell^{-1}$. Salinities between 8 and 12‰ were often associated with similarly hypoxic water. The spatial distribution of the higher bottom salinity observations also closely defined the areas of maximum vertical stratification.

Normally, oxygen-consuming respiratory processes dominate the lower, subeuphotic zone of the water column, while photo-synthetic, oxygen-producing processes prevail in the upper, photic zone of the water column. This is particularly true under density stratified conditions.

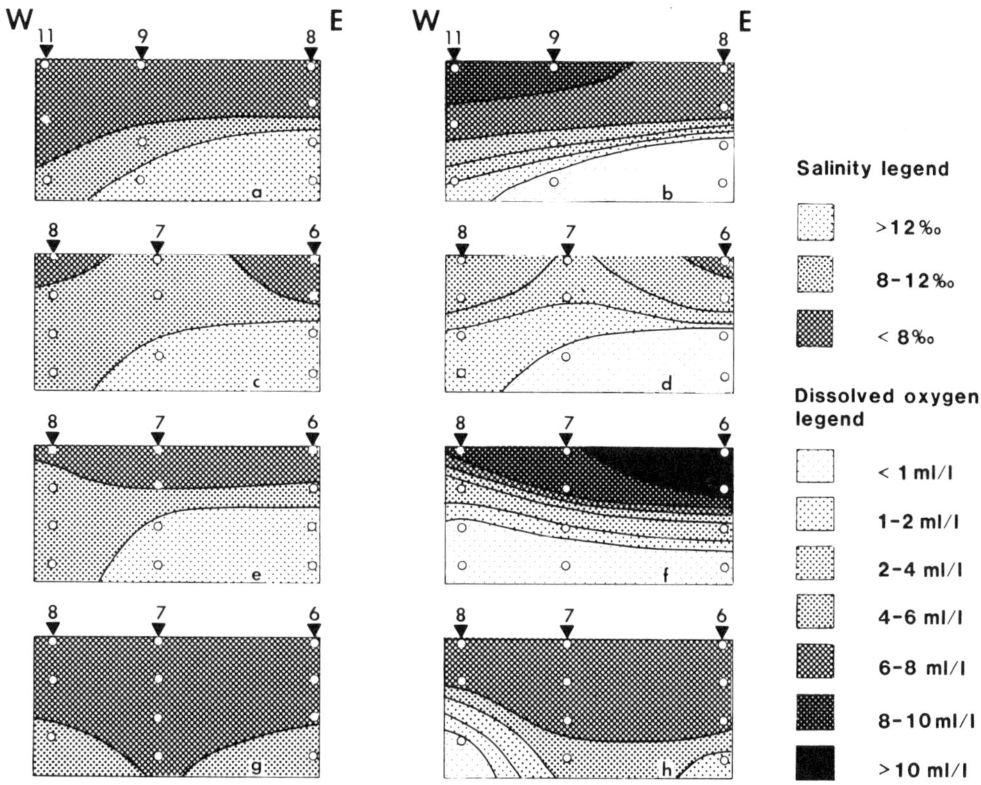

FIGURE 6. Vertical salinity (a, c, e, g) and dissolved oxygen (b, d, f, h) sections corresponding to the lines indicated on Figure 5. Dots represent sample stations.

The cross sections in Figure 6 illustrate the relationship between salinity stratification and the vertical distribution of dissolved oxygen.

The sections in Figure 6 run SE-NW through the northeast part of the bay (see Figure 5). They represent data collected at the same time as that shown in Figure 5. The oxygen distributions in Figures 6b and 6d closely mimic the corresponding salinity patterns of Figures 6a and 6c. These represent both high and low tide conditions. It is clear, though, from the remaining panels that density is not the sole determinant of the dissolved oxygen distribution. The near-surface dissolved oxygen pattern of Figure 6f resembles the salinity pattern in Figure 6e. Both distributions show strong vertical gradients between 1 and 2 m depth. The near-bottom dissolved oxygen distribution, though, does not reflect the significant salinity front between stations 7 and 8. Similarly, the strongest dissolved oxygen gradients in Figure 6h are near bottom around station 8, while the strongest gradients in the corresponding salinity section are near bottom around station 6. Also, some degree of dissolved oxygen stratification may exist even under conditions of a well-mixed water column, as seen in Figures 6g and 6h near station 7.

Diel variations in dissolved oxygen concentrations are also evident in Figure 6. Above the halocline, daytime increases (Figures 6b and 6f) and nighttime decreases (Figures 6d and 6h) in dissolved oxygen concentrations suggest net photosynthetic and net respiratory activity, respectively. Horizontal gradients and the size of the tidal excursion are insufficient to explain the observed changes as the result of purely advective processes. In the lower water column, under stratified and hypoxic conditions, the diel differences in dissolved oxygen concentrations appear dominated by physical processes (e.g., horizontal and vertical advection and diffusion) rather than biological processes.

The relative importance and magnitude of the various oxygen-consuming processes have not yet been definitively determined for Mobile Bay. In the late 1970s, Pamatmat[17] investigated the rates of oxygen consumption in sediment cores from the Bay bottom. He had postulated that, because of the shallowness of the Bay, those processes associated with the bottom and benthos would be relatively more important than those associated with the water column and the total biological activity of plankton and nekton. His findings, based on limited experimental results during the fall, winter, and spring season, are

1. Total oxygen uptake values, at 20°C, ranged between 12.9 and 41.7 mℓ O_2 m^{-2} hr^{-1} and averaged 23.9 mℓ O_2 m^{-2} hr^{-1} (n = 54, s.d. = 5.8 mℓ/m²/hr
2. Chemical oxygen uptake values ranged between 4.8 and 18.1 mℓ O_2 m^{-2} hr^{-1} and averaged 10.9 mℓ O_2 m^{-2} hr^{-1} (n = 42, s.d. = 4.4 mℓ/m²/hr)

The water column vertically homogenizes on subtidal time scales. The stratification, though, rapidly recovers after strong winds die down. It is reasonable to ask whether the time between wind events (10 to 15 days) is sufficient to generate hypoxic conditions considering available estimates of benthic oxygen demand. Pamatmat's estimate was based on a temperature of 20°C. Summer temperatures in the Bay are closer to 28°, which should result in an increased oxygen consumption rate. On the other hand, as oxygen is removed from the lower layer, the benthic respiration rate should decrease. Hargrave[18] presents formulas for benthic consumption within a Canadian lake as a function of temperature and a similar function constructed from a variety of rates taken from the literature. At 28°C, these two formulas suggest rates of 29 and 90 mℓ O_2 m^{-2} hr^{-1}, respectively. For order of magnitude arguments, we will assume a consumption rate of 50 mℓ O_2 m^{-2} hr^{-1} during the summer months. If a 2-m thick layer of water has an initial dissolved oxygen concentration of 8 mℓ $ℓ^{-1}$, then, at this consumption rate, the oxygen will be totally utilized within 13.3 days. These numbers are, clearly, meant to be no more than suggestive. They do, though, indicate that benthic consumption is capable of generating near-bottom hypoxic conditions between summer wind events.

There is little information concerning the areal extent of hypoxic conditions in the Bay. Isolated data points suggest such conditions may occur anywhere within the Bay at any given time, but we know of only one quasisynoptic dissolved oxygen survey of the Bay (Figure 7). This was completed on July 18, 1978. The suggested contours imply an area of 210 km² where near-bottom dissolved oxygen content was below 1 mℓ $ℓ^{-1}$ and 365 km² where it was below 2 mℓ $ℓ^{-1}$. Major parts of the Bay bottom may be simultaneously hypoxic.

V. CONCLUSIONS

During summer months, strong haline stratification isolates the bottom waters of Mobile Bay from direct air-sea interaction. High temperatures increase metabolic rates and benthic consumption reduces near-bottom dissolved oxygen content of the water column to values that are stressful to the biota. This hypoxic water is advected by tides and wind-driven baroclinic motions. This movement appears to be responsible for the jubilee phenomenon in the Bay.

Large areas of the Bay are affected by hypoxic conditions. Strong wind, though, can rapidly dissipate these conditions. Hypoxia quickly redevelops following reestablishment of the stratification after a wind event.

Hypoxic conditions are not generally found in winter. Presumably, this is because wind mixing is more frequent than in summer and lower temperatures result in reduced benthic oxygen consumption rates. Oxygen depletion has been reported during periods of spring river flooding episodes, but no "jubilee type" activities have been observed during these events.

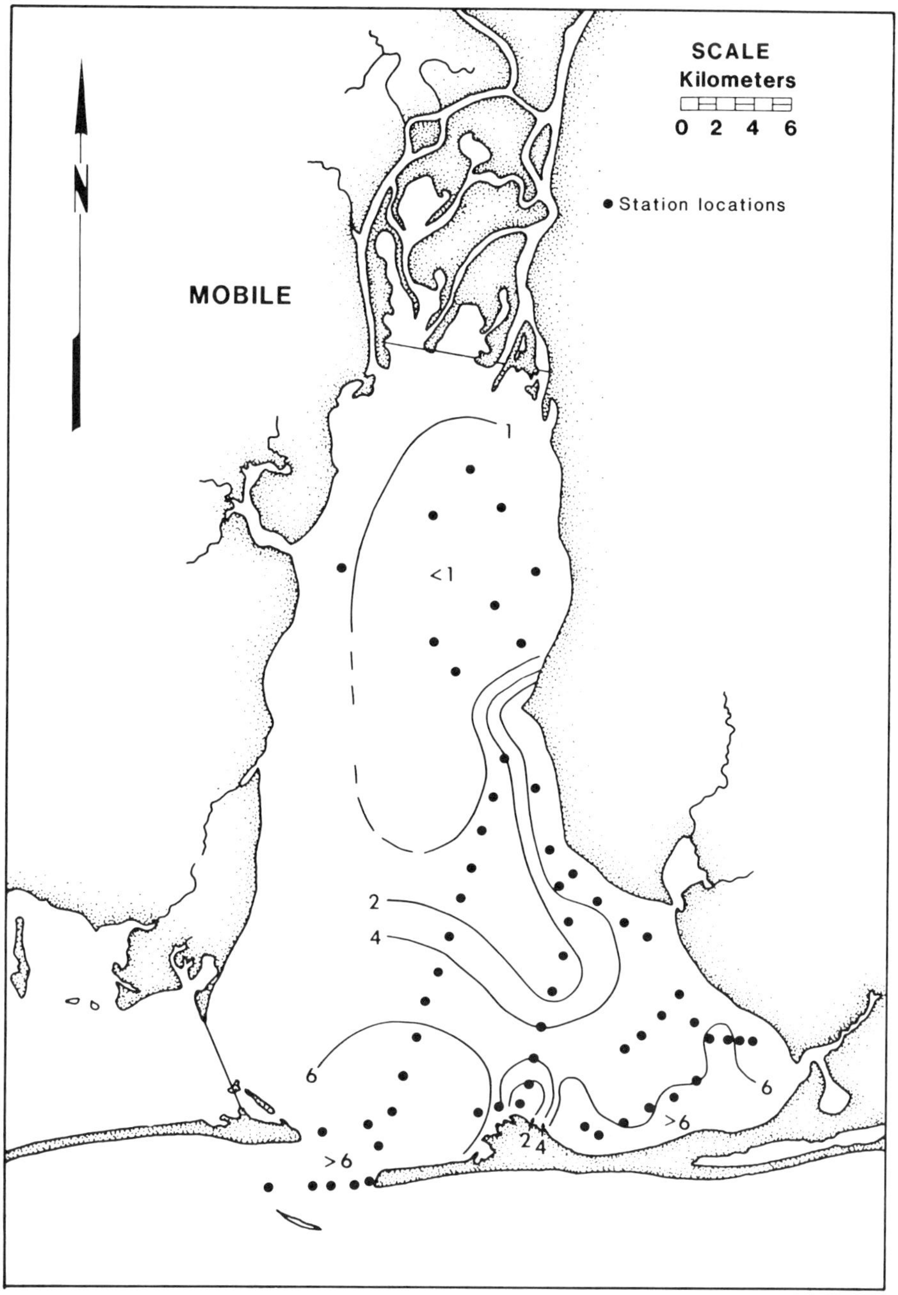

FIGURE 7. Bottom dissolved oxygen field on July 18, 1978.[21] Dots indicate station locations. Contours are in mℓ/ℓ.

ACKNOWLEDGMENTS

Funding for this work was provided by the Louisiana Sea Grant University Program, a part of the National Oceanic and Atmospheric Administration of the U.S. Department of Commerce, the Louisiana State University Department of Marine Sciences, the University of Alabama Marine Science Program, and the Dauphin Island Sea Lab. The figures were prepared by Ms. C. Harrod.

(Marine Environmental Sciences Consortium Contribution No. 81 and Contribution No. 78 from the Aquatic Biology Program of the University of Alabama.)

REFERENCES

1. **Chow, V. T.,** *Handbook of Applied Hydrology,* McGraw-Hill, New York, 1964, 1467.
2. **Schroeder, W. W.,** Riverine influence on estuaries: a case study, in *Estuarine Interactions,* Wiley, M. L., Ed., Academic Press, New York, 1978, 347.
3. **Schroeder, W. W.,** The dispersion and impact of Mobile River system waters in Mobile Bay, Alabama, WRRI Bulletin 37, Water Resources Research Institute, Auburn University, Auburn, Ala., 1979, 48.
4. **Peirce, L. B.,** Surface water in southwestern Alabama. *Ala. Geol. Sur. Bull.,* 84, 1, 1966.
5. **Schroeder, W. W.,** unpublished data, 1980.
6. **Austin, G. B., Jr.,** On the circulation and tidal flushing of Mobile Bay, Alabama, Part 1, Technical Report No. 12, Oceanographic Survey of the Gulf of Mexico, Texas A and M Research Foundation, College Station, Texas, 1954, 28.
7. **Schroeder, W. W. and Lysinger, W. R.,** Hydrography and circulation of Mobile Bay, in *Symposium on the Natural Resources of the Mobile Bay Estuary,* Loyacano, H. A. and Smith, J. P., Eds., U.S. Army Corps of Engineers, Mobile District, Mobile, Alabama, 1979, 75.
8. **Bondar, C.,** Contributie la Studiul Hidraulic al Iesirii la Mare prin Gurile Dunarii, *Studii de Hidrologie,* XXXII (Probleme de Oceanografie), Institutul de Meteorologie si Hidrologie, Bucuresti, 1972, 466.
9. **Ling, T.-F. T.,** Hydrodynamics of Mobile Bay and Mississippi Sound, *Bureau of Engineering Research Report No. 283-112,* University of Alabama, University, 1981, 149.
10. **April, G. C., Raney, D. C., Chern, L., Jarrel, J. P., Lau, D.-J., and Wu, Y.-C.,** Hydrodynamics of Mobile Bay and Mississippi Sound, *Bureau of Engineering Research Report No. 247-112,* University of Alabama, University, 1980, 98.
11. **Youngblood, J. N. and Raney, D. C.,** Salinity propagation in Mobile Bay, *Proceedings, IASTED International Symposium,* ASM 1983, Applied Simulation and Modelling, San Francisco, May 16—18, 1983, ACTA Press, Anaheim, California, 1984, 16.
12. **Loesch, H.,** Sporadic mass shoreward migrations of demersal fish and crustaceans in Mobile Bay, Alabama, *Ecology,* 41, 292, 1960.
13. **May, E. B.,** Extensive oxygen depletion in Mobile Bay, Alabama, *Limnol. Oceanogr.,* 18, 353, 1973.
14. **Schroeder, W. W.,** The impact of the 1973 flooding of the Mobile River system on the hydrography of Mobile Bay and east Mississippi Sound, *Northeast Gulf Sci.,* 1, 68, 1977.
15. Marine Environmental Sciences Consortium, Biological baseline studies of Mobile Bay, May, 1980, Dauphin Island Sea Lab Interim Technical Report III, Dauphin Island, Alabama, 1980, 180.
16. **Tatum, W.,** personal communication, 1978.
17. **Pamatmat, M.,** personal communication, 1980.
18. **Hargrave, B. T.,** Similarity of oxygen uptake by benthic communities, *Limnol. Oceanogr.,* 14, 801, 1969.
19. Marine Environmental Sciences Consortium, Biological baseline studies of Mobile Bay, Alabama, August 1980, Dauphin Island Sea Lab Interim Technical Report VI, Dauphin Island, Alabama, 1980, 215.
20. Marine Environmental Sciences Consortium, Analysis of an environmental monitoring program Theodore Ship Channel and barge channel extension, Mobile Bay, Alabama. Dauphin Island Sea Lab Technical Report No. 83-003, Dauphin Island, Alabama, 1983, 106.
21. **Schroeder, W. W.,** The dissolved oxygen puzzle of the Mobile Estuary, in *Symposium on the Natural Resources of the Mobile Bay Estuary,* Loyacano, H. A. and Smith, J. P., Eds., U.S. Army Corps of Engineers, Mobile District, Mobile, Alabama, 1979, 25.

Chapter 14

CIRCULATION ANOMALIES IN TROPICAL AUSTRALIAN ESTUARIES

Eric Wolanski

TABLE OF CONTENTS

I. INTRODUCTION

Tropical estuaries of Northern Australia are characterized by large seasonal fluctuations in the freshwater discharge. From January to March the "wet season" prevails, during which intense rainfalls can occur, but these are very irregular. The freshwater discharge is usually quite "small" except during short-lived floods. After scattered rainfalls in the austral winter, which usually lead to only minor runoffs, hot, dry weather prevails from September to November during which time the cloud coverage is usually minimal, rainfall is nonexistent, wind is slight, and the air temperature frequently exceeds 30°C. The freshwater discharge becomes quite small, of the order of a few m^3 sec^{-1} for major rivers, though it seldom vanishes.

The estuaries respond to these seasonal fluctuations. Because the wet weather situation is most spectacular and obvious, studies by physical oceanographers have for a long time been restricted to the fate of flood waters in shelf seas. It was shown that river floods are formed in the coastal ocean, but, because of their transient nature they are very patchy and short lived.[1,2] The recovery of the estuaries after floods has not been studied.

Very recently, some attention has been given to the dry season estuarine dynamics. So far however, the attention given by physical oceanographers to northeastern tropical Australian estuaries in dry weather has been limited to five estuaries. These water bodies, shown in Figure 1, are the Alligator River — Van Diemen Gulf system, the Norman River, the Ducie-Wenlock River, the Escape River, and Coral Creek. The results show the importance of evaporation and evapotranspiration in driving the baroclinic estuarine circulation. These findings are sketched below, starting with a description of the Alligator River-Van Diemen Gulf system.

II. THE ALLIGATOR RIVER — VAN DIEMEN GULF

A map is given in Figure 2a. The South Alligator River is tidal for about 100 km. It is 10 m deep in the thalweg at high tide, with a strong tidal current (1 m sec^{-1} at averaged tides). The Van Diemen Gulf is shallow, less than 20 m deep up to 50 km offshore. It is connected to the Arafura Sea and the Indian Ocean through the deep Dundas Strait and also through the much narrower, shallower, and reef-studded Clarence Strait.

Vertical profiles of temperature (T) and salinity (S) were obtained at a number of stations shown in Figure 2a, and the data are shown both as typical vertical profiles (Figure 2b) and as T-S diagrams (Figure 2c). It can be seen that the South Alligator River waters were quite well-mixed vertically. At 35 km from the river mouth, the temperature was about 29.4°C and the salinity 32.9‰ (station 1). The salinity increased with distance downstream. There was evidence of warm Gulf water of greater salinity intruding up-river near the bottom (station 4). In the southern region of the Van Diemen Gulf (stations 5 to 17), a 2 to 3 m thick surface layer of very warm and saline water ($T \gtrsim 30°C$, $S = 35.6‰$) was present, floating over the slightly more saline (by 0.2‰) and colder (by 1°C) bottom waters (e.g., see the profile for Station 5). Occasionally, a 2-m thick third layer was found near the bottom, consisting of more saline (by 0.6‰) and warmer (by 0.2°C) waters (e.g., station 8). The fact that both temperature and salinity are larger at the bottom may generate salt fingering.

Both salinity and temperature decreased regularly across the Gulf toward the Arafura Sea which was strongly temperature stratified (e.g., station 25).

Thus, the Van Diemen Gulf, up to 30 km from the shore, is a zone of maximum salinity. The T-S diagram (Figure 2c) shows the existence of three distinct water masses. These are the estuarine waters, the salinity maximum zone waters, and the Arafura Sea waters. The profiles taken at sites located in between these respective water masses, have T-S values

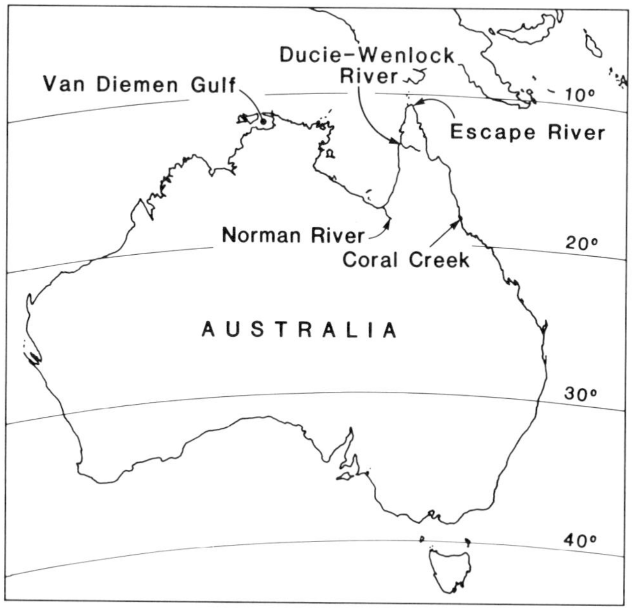

FIGURE 1. General location map.

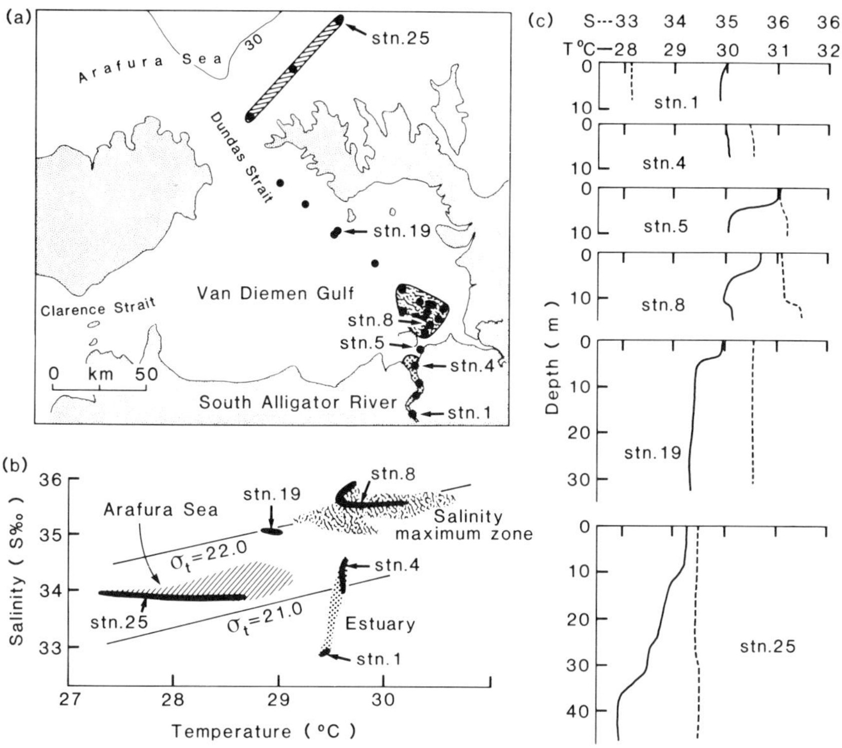

FIGURE 2. Part (a) is a map of the South Alligator River-Van Diemen Gulf system, with depth in fathoms (1 fathom = 1.8 m), and location of profiling stations. In (b) are shown selected vertical profiles of temperature and salinity. In (c) is shown a temperature-salinity (T-S) diagram, with the shading types corresponding to the areas similarly shaded in (a). The T-S data were obtained in October, 1983.

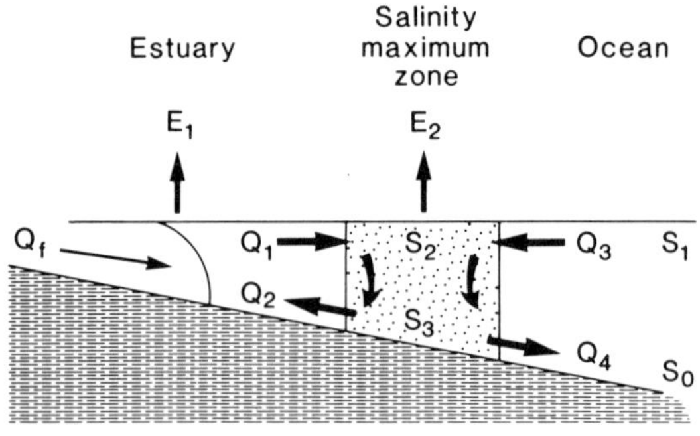

FIGURE 3. Sketch of the internal estuarine circulation.

intermediate between those of the surrounding two primary water masses, suggesting that the intermediate water masses result from mixing of the two surrounding primary water masses. For instance, the T-S data at site 19 fall in between those of the Arafura Sea and of the salinity maximum zone (Figure 2c).

Thus, it appears that there is no direct mixing and exchange of water between the estuary and the Arafura Sea and that the exchange can only occur through the salinity maximum zone.

Evaporation is presumed to be the reason for this unusual estuarine behavior. A box model of the internal circulation is sketched in Figure 3. The system is idealized by three water masses: first, the estuary (the South and East Alligator rivers) receiving freshwater at a rate Q_f and losing water by evaporation at a rate E_1; second, the salinity maximum zone (the upper waters of the Van Diemen Gulf) losing water at a rate E_2; and third, the ocean. Denoting the respective salinities and discharges of the top and bottom layers by S_i and Q_i (see Figure 3), the equations for conservation of water and salt in the estuary are

$$Q_f + Q_2 = Q_1 + E_1 \quad (1)$$

$$Q_1S_2 = Q_2S_3 \quad (2)$$

and those for the estuary and the salinity maximum zone are

$$Q_f + Q_3 = Q_4 + E_1 + E_2 \quad (3)$$

$$Q_3 S_1 = Q_4 S_0 \quad (4)$$

Hence, from Equations 1 to 4,

$$Q_3 = (E_1 + E_2 - Q_f)/(1 - S_1/S_0) \quad (5)$$

and

$$Q_1 - (Q_f - E_1)/(1 - S_2/S_3) \quad (6)$$

Assuming that free-water evaporation occurs over the surface area of the respective water bodies, one estimates $E_1 \cong 2\ m^3\ sec^{-1}$, and $E_2 \cong 250\ m^3\ sec^{-1}$. For typical values of the S_i, and for $Q_f = 10\ m^3\ sec^{-1}$, one finds $Q_1 = 2 \times 10^2\ m^3\ sec^{-1}$, and $Q_3 = 2 \times 10^4\ m^3\ sec^{-1}$. The flushing time of the Van Diemen Gulf waters is estimated to be about 100 days, so that, in view of weather conditions, a quasisteady state situation prevailed during the cruise.

Hence, since $Q_1 > 0$ and $Q_3 > 0$, as sketched in Figure 3, a classical internal estuarine circulation prevailed in the South Alligator River, but an inverse estuarine circulation prevailed in the northern region of the Van Diemen Gulf. To conserve mass, downwelling occurred in the southern regions of the Van Diemen Gulf, i.e., the salinity maximum zone. Because the width of the salinity maximum zone (about 30 km) is larger than the tidal excursion ($\leq$ 10 km), mixing between Arafura Sea waters and South Alligator River waters was strongly inhibited. In fact, in the dry season, the South Alligator River waters are essentially trapped in the estuary. The ecological implications of this unusual estuarine circulation may be profound.

III. THE NORMAN RIVER

Recently, in November 1985, a study was undertaken in the Norman River (Figure 1), an estuary roughly 80 km long, located in the southern region of the Gulf of Carpentaria (latitude 17.5°S). The tides are diurnal with maximum amplitude of 3.6 m. The estuary is surrounded by vegetation-free salt pans, several kilometers wide on both banks. These salt pans are only inundated by the tides a few times a year, remaining dry for about 6 months of the year (in winter) when mean sea level is about 0.5 m lower than in summer. The field study was timed to coincide with the first tidal inundation of the salt pans in summer. Before tidal inundation, a salinity maximum zone was found in the estuary, 30 km long, with the peak salinity excess being about 2‰. In such conditions, freshwater never reaches the Gulf of Carpentaria. Following tidal inundation of the salt pans, this salinity excess increased in 1 day to 3‰. This additional salt was carried into the estuary by the water returning to the estuary after flooding the salt pans (to a depth of typically 10 cm). The salinity of this water reached 42‰, compared to typically 36 to 37‰ at rising tide.

Hence, while free water evaporation creates a salinity maximum zone in the Norman River, evaporation in the adjoining salt pans results in occasional additions of salt to the estuary, thereby reinforcing the salinity maximum zone.

IV. EVAPOTRANSPIRATION

A salinity maximum zone was also found in the other tropical estuaries shown in Figure 1 that have been studied during the dry season. In these cases, the salinity maximum zone was located in the estuary just upstream of the river mouth. Hence, an inverse estuarine circulation appears to be present in the dry season near the river mouth in all the Australian tropical estuaries that have been studied. At first, this may appear surprising because the free-water surface area of the salinity maximum zone is two orders of magnitude smaller than in the Van Diemen Gulf. However, contrary to the situation at the South Alligator River, and the Norman River, these rivers are fringed by extensive areas of mud banks and densely vegetated mangrove swamps. These extensive intertidal areas generate, by evaporation and evapotranspiration, a large water loss, the salt being left behind and brought back into the river at the following tidal cycle. Further, it is believed that the evaporation rate from the mud banks exposed at low tides, may be orders of magnitude higher than the free water one, because of the high mud temperature (up to 50°C).

Evapotranspiration can also play an important role. Indeed, Wolanski et al.[3] and Wolanski and Gardiner[4] have shown that evapotranspiration from mangrove swamps can, in the absence of freshwater runoff, drive an inverse estuarine circulation, in the creek draining the swamps. The study site was Coral Creek, a 6-km-long tidal creek-mangrove swamp system sketched in Figure 4a. While the creek is deep enough to never dry out at low tide, the surrounding mangrove swamps are almost completely immersed at 1-m tidal elevation and have no surface water at 0-m tidal elevation. Figure 4c shows the longitudinal distribution of salinity

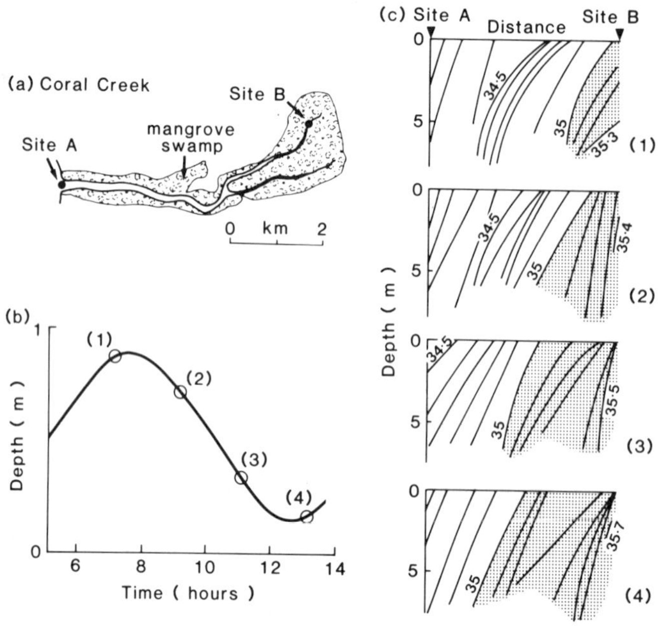

FIGURE 4. In (a) is shown a map of Coral Creek, a tidal creek-mangrove swamp system. In (b) is shown the observed tidal elevation during a dry weather study, and in (c) are shown successive plots of the longitudinal distribution of salinity in the creek at the times shown in (b). The contour interval for salinity in (c) is 0.1. The data were collected in September, 1979.

in the creek at successive times during the falling tide shown in Figure 4b. It can be seen that the estuary was partially mixed. The salinity decreased toward the mouth of the estuary, and the shape of the isohalines, inclined downward toward the estuary mouth, suggests the presence of an inverse estuarine circulation. Also, as the swamps were draining into the creek at falling tide, a pool of high salinity swamp water (S > 35‰) was increasing the salinity of the water in the creek. These high salinity swamp waters were pushed back into the swamps at rising tide.

Since evapotranspiration occurs all the time and since there is no evidence of an accumulation of salt in the mangrove swamps, a salinity balance must prevail after averaging over a spring-neap tide cycle. As a result, the mean export of salt away from the creek by advection and diffusion (enhanced by the inverse estuarine circulation) equals the mean import of salt into the creek from the swamps after, by evapotranspiration, the mangrove trees take the water and leave the salt behind. The resulting inverse estuarine circulation implies a long residence time in the absence of freshwater runoff, and is presumed to be responsible for the occasional occurrence of anaerobic conditions in bottom waters in the upper parts of the creek.

REFERENCES

1. **Wolanski, E. and Jones, M.,** Physical properties of Great Barrier Reef lagoon waters near Townsville. I. Effects of Burdekin River floods, *Aust. J. Mar. Freshwater Res.*, 32, 305, 1981.
2. **Wolanski, E. and van Senden, D.,** Mixing of Burdekin River flood waters in the Great Barrier Reef, *Aust. J. Mar. Freshwater Res.*, 34, 44, 1983.

3. **Wolanski, E., Jones, M., and Bunt, J. S.,** Hydrodynamics of a tidal creek-mangrove swamp system, *Aust. J. Mar. Freshwater Res.*, 31, 431, 1980.
4. **Wolanski, E. and Gardiner, R.,** Flushing of salt from mangrove swamps, *Aust. J. Mar. Freshwater Res.*, 32, 681, 1981.

Chapter 15

PHYSICAL OCEANOGRAPHY OF THE ST. LAWRENCE ESTUARY*

Mohammed I. El-Sabh

TABLE OF CONTENTS

* Contribution du Laboratoire Océanologique de Rimouski

ABSTRACT*

In recent years major scientific and intensive field programs were carried out in the St. Lawrence estuary to obtain a better understanding of the circulation dynamics and distribution of properties and to place it in a better perspective relative to other major estuaries of the world. This paper reviews our state of knowledge and describes the tidal, seasonal, and annual variability of freshwater, currents, and water properties of the estuary. In this regard, some conclusions of general interest are as follows:

1. All three types of estuarine mixing exist in the St. Lawrence estuary: the well-mixed, at the head of the estuary; the moderately-mixed, in the section below the head; and the stratified in the lower part.
2. Both advection and diffusion contribute importantly to the upstream salt flux in the upper part while advective processes account for more than 99% of upstream salt transfer in the lower part.
3. Because of the presence of internal waves, fronts, and extensive tidal mixing, vertical oscillations of up to 80 m of the intermediate cold layer are common in the area near the Saguenay entrance.
4. The lower estuary is characterized by two circulation modes: the first is a series of gyres with alternating rotational senses along the estuary accompanied with transverse currents and fronts between any two gyres, and the second mode shows the estuary divided longitudinally into two halves, the southern one characterized by cold, more saline, and denser waters compared to the northern part. The combined actions of the neap-spring tidal cycle, meteorological forcing, freshwater discharge, and topography are believed to be the cause for such variability.

I. INTRODUCTION

The St. Lawrence estuary in eastern Canada (Figure 1) is the region where the waters of the Atlantic Ocean and those of the Great Lakes co-mingle. It is 500 km from the Great Lakes and nearly 1000 from the ocean and is a middle ground generally characterized by waters abundant in life. Apart from its three-dimensional characteristics, large size, and topographic features, the St. Lawrence estuary provides a unique case to study from two other standpoints. First, the formation of ice in the winter months affects the wind-induced circulation and mixing. Ice formation starts in December, and there is almost a complete ice cover in the estuary from about January to March.[1] In addition to the effect of reducing wind-induced mixing and circulation, the ice cover provides another boundary layer which increases the effects of friction and the accompanying generation of turbulent mixing processes. Furthermore, the freezing and melting of sea ice must be considered, as the formation of ice removes freshwater from the water column and releases it upon melting. Although the total annual discharge from this process is zero if the advection of ice can be neglected, the equivalent discharge is at times significant. Using data given by Forrester and Vandall,[2] Budgen et al.[3] estimated the discharge implied by ice freezing as -8×10^3 m^3/sec in mid-February and of the same magnitude, but positive, in mid-April. As will be seen later, this is a significant effect compared to the river discharge from the St. Lawrence and major tributaries.

The second characteristic of the St. Lawrence estuary is that the flow of many rivers entering it are subject to extensive alteration, principally for hydroelectric power generation. Modifying the natural seasonal runoff for human convenience interferes with the hydrological

* Key words: Variability, mixing, gyres, fronts, upwelling, currents, temperature, and salinity, St. Lawrence estuary.

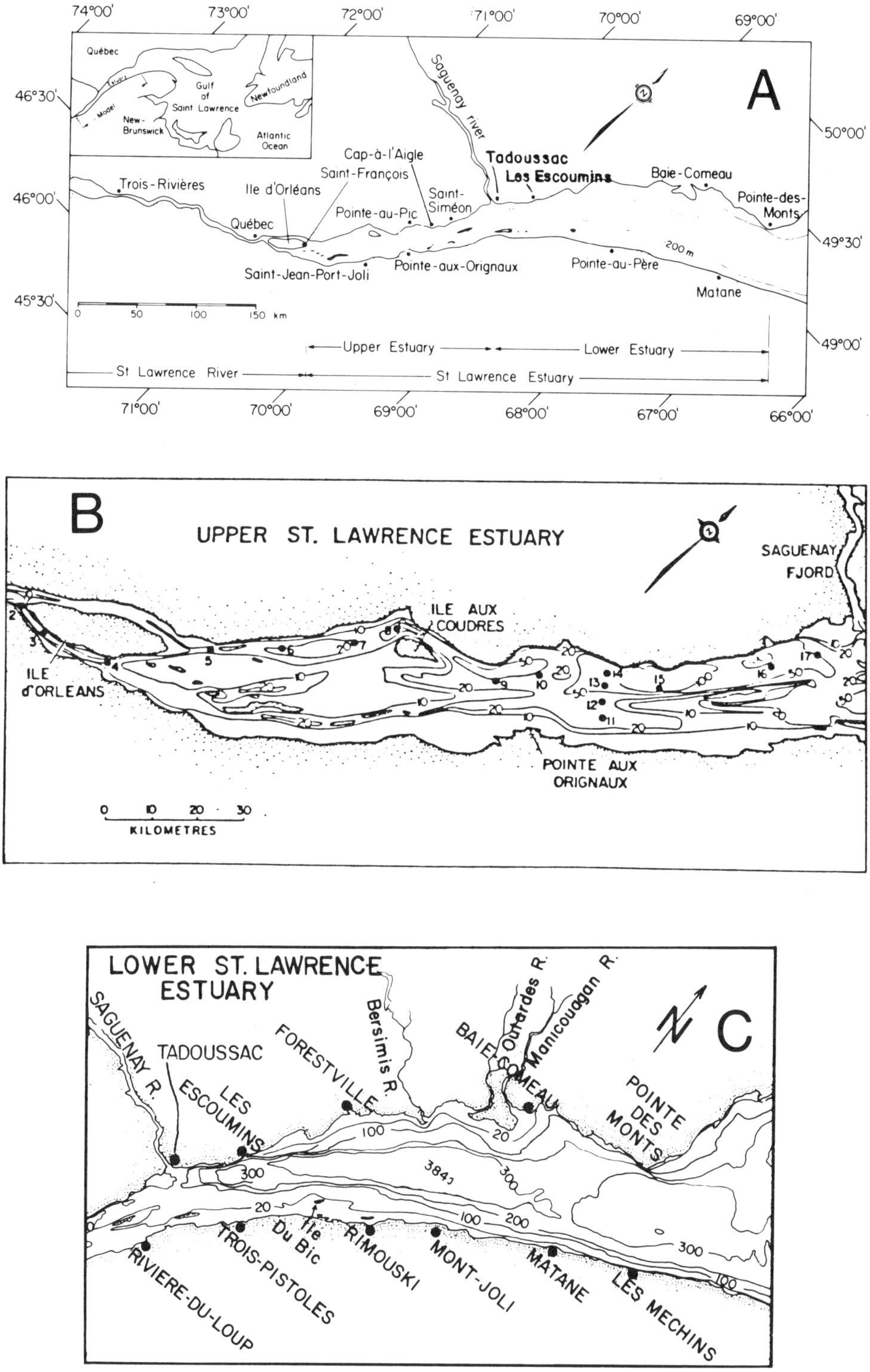

FIGURE 1. The St. Lawrence estuary (Canada). Bathymetry is shown in meters.

cycle and with the physical and biological balance of the estuary and adjacent coastal region.[4,5] Under these circumstances, it becomes a matter of particular interest to understand, predict, and control the consequences of these man-made modifications.[6] The present paper is a continuation of a review published by El-Sabh.[1] It is an attempt to summarize our present state of knowledge on the hydrodynamics of the St. Lawrence estuary and enable one to place it in a better perspective relative to other major estuaries of the world.

II. PHYSIOGRAPHIC SETTING

The area between Trois-Rivières and Québec City (Figure 1) represents the tidal portion of the St. Lawrence River. The St. Lawrence estuary, with an area of 10,800 km^2, begins at the landward limit of the saltwater intrusion near Ile d'Orléans and extends about 400 km seaward to Pointe-des-Monts, where its channel opens into the Gulf of St. Lawrence. The estuary is commonly subdivided into its upper and lower regions by a transverse section at Tadoussac, near the mouth of the Saguenay fjord. This is a convenient way in which to bisect the estuary since the bathymetry and water circulation differ markedly in each half.

The upper estuary (Figure 1b) varies in width from 2 km near Ile d'Orléans to 24 km near Tadoussac. It has an uneven bottom topography with several disconnected channels and troughs separated by ridges and islands. The southern side is shallower and is characterized by depths of less than 10 m. The northern side has a nearly continuous channel with depths increasing from about 10 m in the shoals near Ile d'Orléans to 120 m in the deep basin at the north-eastern extremity. In the lower estuary (Figure 1c), between Tadoussac and Pointe-des-Monts, the beginning of the Laurentian Channel forms a deep central trough with depths greater than 300 m. Typical width and cross-sectional area at the mouth are 50 and 11 km^2, respectively. Its physiography, particularly the depth of the basin, makes it more typical of continental-shelf environments than of other shallower estuaries.

III. ESTUARINE CLASSIFICATION

The St. Lawrence estuary shows several types and different mixing conditions along its longitudinal axis. On the basis of Pritchard's[7] classification, the extent of mixing in the estuary varies from nearly vertically homogeneous in the vicinity of Ile d'Orléans, to moderately mixed in most of the upper estuary, to stratified in the lower estuary.[8] Fjord-like conditions can also be found at the entrance to the Saguenay River, near Tadoussac, where a shallow sill of 20 m separates the deep water of the St. Lawrence estuary from that of the Saguenay fjord.

Hansen and Rattray[9] proposed an estuarine classification scheme in which estuarine dynamics are explicitly included. Using all available data, the two dimensionless theoretical parameters of Hansen and Rattray were calculated by El-Sabh[1] for seven transects in the lower estuary and Gulf of St. Lawrence and by Meric[10] for two transects in the upper estuary. These results are plotted on the stratification-circulation diagram and can be compared with other estuaries (Figure 2a). The positions of the nine transects are shown in the inset of Figure 2a. All upper St. Lawrence estuary data collected between Ile-aux-Coudres and Rivière-du-Loup (lines I_c, I_o, and BB), fall in the type 2b estuary for which stratification is classed as "appreciable". In this region, diffusive processes account for more than 40% of the fluxes. Data for the region between Ile-du-Bic and Pointe-des-Monts of the lower estuary (lines DD, EE, FF, and GG), together with Cabot Strait (line CS) in the Gulf of St. Lawrence, fall into the area of the diagram corresponding to type 3b. This describes a partially mixed estuary in which advective processes account for more than 99% of the upstream salt transfer. This also implies that the gravitational circulation reaches its maximum development there. At transect CC, between Ile-du-Bic and the head of the Laurentian trough near Tadoussac,

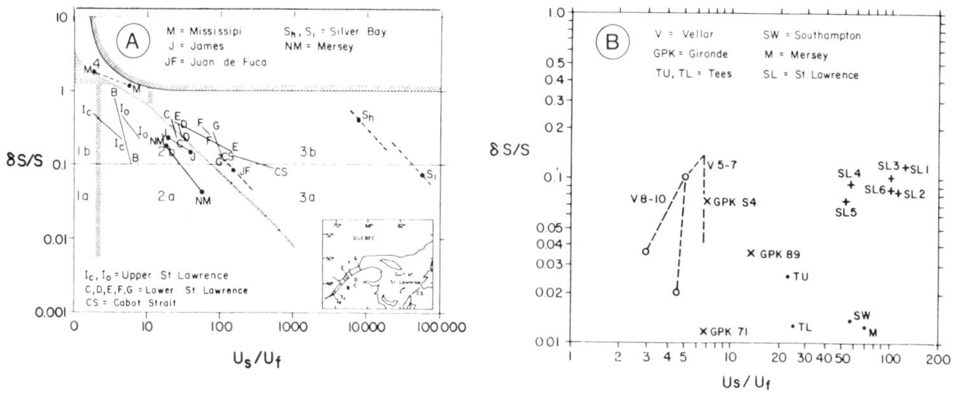

FIGURE 2. A) Longitudinal and B) lateral estuarine classification of the St. Lawrence estuary.[1,12]

the estuary is transitional between types 3b and 2b, meaning that diffusive processes begin to account for more than 10% of the property fluxes. Topographic features in this region, such as a wide shallow plateau along the southern part and a bottom rise from deeper than 340 m to less than 40 m over a distance of only 20 km, are believed to be of major importance.

In the estuarine classification scheme outlined above, there is the implicit assumption that the circulation occurs entirely in the longitudinal sense, along the estuary, and that there are no differences in the velocity or density fields across the estuary. This does not appear to be strictly valid for many cases,[11] and certainly not for the St. Lawrence estuary. Thus, because of the cross-sectional averaging required by Hansen and Rattray's classification method, the above estimated diffusive flux is probably a function of lateral circulation rather than turbulent mixing. Lateral stratification at the Matane cross section, near the mouth of the St. Lawrence estuary, was studied by Larouche[12] during the high spring runoff of 1979. When the lower St. Lawrence estuary is compared with other estuaries on the lateral stratification-circulation diagram (Figure 2b), it is classified as the most stratified.

IV. FRESHWATER DISCHARGE

Freshwater inflow enters the estuary mainly from the St. Lawrence river and its tributaries, including the Saguenay River and the group of rivers consisting of the Bersimis, the Outards, and the Manicouagan (henceforth abbreviated as B.O.M. rivers). These major tributaries enter the system from the north shore of the estuary, while inflow from the south shore is insignificant.

Available information on the monthly mean discharge from all rivers around the St. Lawrence estuary for the period 1949 to 1981 were used to estimate the seasonal and long-term fluctuations. With a mean discharge of 11,900 m^3/sec at Québec City, the St. Lawrence has the greatest discharge of any North American river draining to the Atlantic coast. Furthermore, as shown in Table 1, the St. Lawrence River provides 80% of the total freshwater runoff into the estuary, while the Saguenay and the north shore (B.O.M) rivers provide only 9.9 and 9.7%, respectively. Notice that since most of these rivers are regulated by man, maximum freshwater discharge does not reach the estuary from all rivers simultaneously. A series of dams installed on the Manicouagan and Outardes rivers since 1964 has shifted much of the spring melt water to summer discharge and the autumn peak rainfalls to winter discharge.[5]

The seasonal character of freshwater discharge at the mouth of the estuary, near Pointe-des-Monts reveals marked monthly fluctuations.[1,3] Freshwater discharge varies from a mean monthly minimum of 13.3×10^3 m^3/sec in February to March to a maximum of 23.4 ×

Table 1
MONTHLY MEAN (10^3 m^3/SEC) RIVER DISCHARGE IN THE ST. LAWRENCE ESTUARY

River	Jan.	Feb.	Mar.	Apr.	May	June	July	Aug.	Sept.	Oct.	Nov.	Dec.	Annual mean
St. Lawrence at Quebec City (1949—1981)	10.47	10.45	11.55	17.01	15.82	12.40	11.12	10.53	10.35	10.80	11.10	10.96	11.90
Saguenay (1949—1981)	0.96	0.96	0.97	1.14	2.80	2.50	1.60	1.42	1.38	1.55	1.36	1.04	1.47
North Shore (B.O.M.) (1964—1981)	1.60	1.61	1.41	1.17	1.62	1.67	1.23	1.28	1.11	1.36	1.46	1.76	1.44

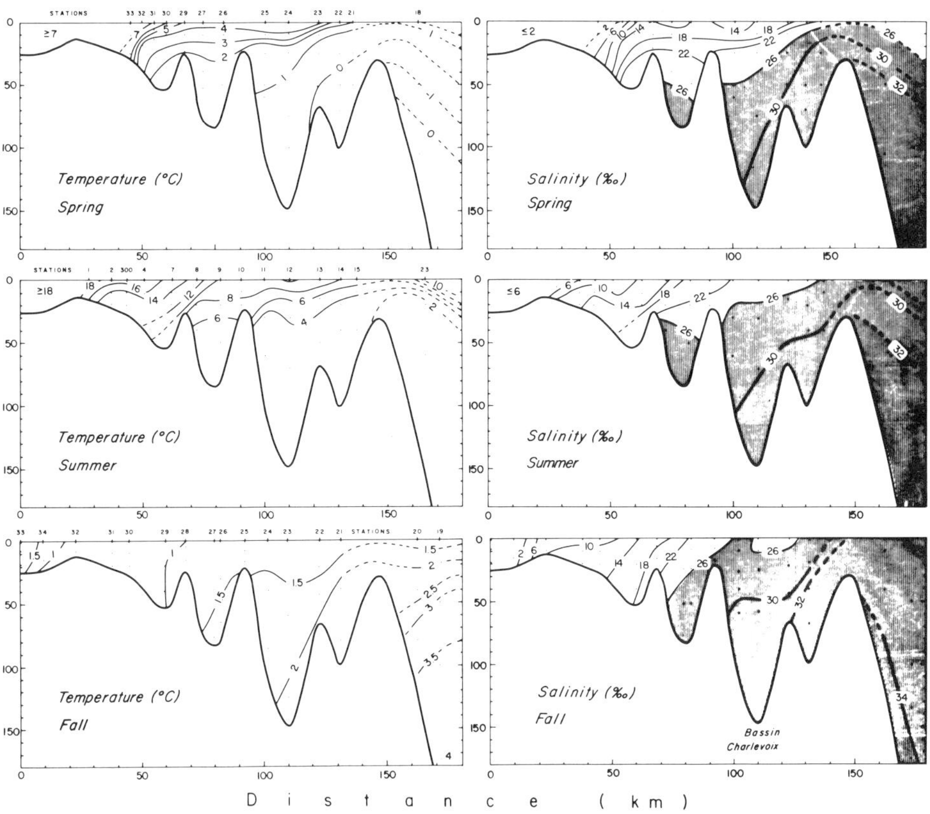

FIGURE 3. Vertical salinity and temperature distributions in the upper St. Lawrence estuary in 1976.[17]

10^3 m^3/sec in May, with a mean annual value of 14.8 × 10^3 m^3/sec. The data show that 26% of the yearly drainage is usually produced during May and June, the two flood months of the year. Furthermore, very large monthly differences occur from year to year due to climatic factors and/or artificial control of the drainage.[1,4] The long-term trend of the freshwater discharge at Pointe-des-Monts shows large cyclical variations with periods on the order of 10 to 11 years.[1,5] Higher discharge occurred in the early 1950s and mid-1970s and lower flows occurred in the middle of the 1960s. Such variations are related to the long-term water level cycles (11 and 22 years) of the Great Lakes.[5]

V. SALINITY AND TEMPERATURE DISTRIBUTIONS

The upstream limit of salt intrusion in the St. Lawrence estuary during the spring freshet occurs on the northern side between Ile d'Orléans and Ile-aux-Coudres, but due to bathymetric and geostrophic effects, salt does not intrude as far upstream on the southern side. Examination of a salinity cross section reveals that the main stream of freshwater flows along the south shore while the compensating flow of saltwater is concentrated in the northern deep channel.[8]

One of the important oceanographic features of the upper estuary, as shown from the longitudinal distributions of salinity and temperature (Figure 3), is the weak stratification that can be observed all year long. The extent of mixing varies from nearly vertically homogeneous at the head of the estuary, to moderately stratified at the eastern end. Variable water flow throughout the year does not alter the character of this distribution. Superimposed on the seasonal variations is a short-term fluctuation due to tides. The tidal movements,

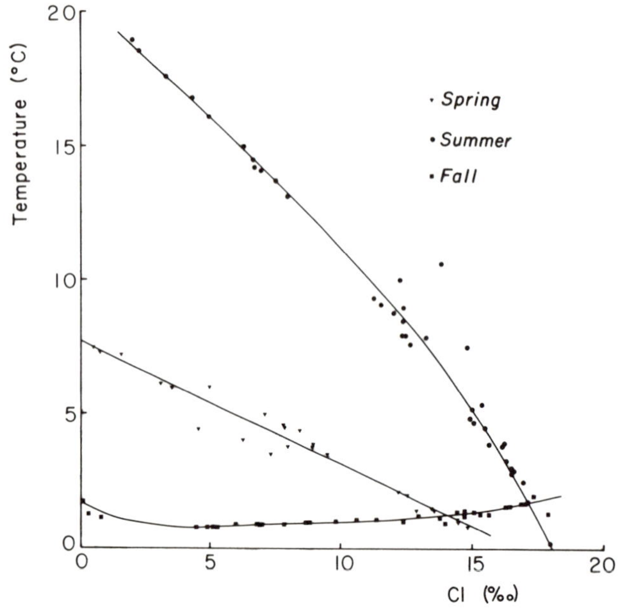

FIGURE 4. Temperature-chlorinity relationships in the upper St. Lawrence estuary in 1976.[17]

with an average amplitude of 5 m between Ile d'Orléans and the Ile-aux-Coudres,[14] bring about a reversible tidal current which can reach a speed of 7 to 8 kn at some places and can move the isohalines several kilometers between high and low water.[15] Muir[16] examined more than 15,000 temperature-salinity (T-S) pairs distributed over 62 13-hr profiling stations. He showed that although the T-S relationship at an individual station is linear and reflects conservative mixing processes, there is no single T-S relationship that will hold over the whole estuary. This is probably due to variability in the end members, which are the St. Lawrence river and the lower estuary. Figure 4 shows that the temperature-chlorinity relationships are almost linear for the ice-free seasons. According to Pelletier and Lebel,[17] this observation is an indication that mixing occurs quickly and that the mixture does not remain in the upper estuary very long.[18] The temperature of the water during the mixing is hardly influenced by the air temperature, but it is primarily influenced by the water masses and the temperature of the surrounding fresh- and seawater.

Based on current meter and CTD observations made in the spring and summer of 1973 in the area between Trois-Pistoles and Rivière-du-Loup, Reid[18] offered the most comprehensive view of the mixing mechanism in the St. Lawrence estuary. He showed that major mixing takes place in an area of abrupt shoaling at the point of confluence of the Saguenay and St. Lawrence due to the combination of internal waves and tidal currents across sills. Fronts which were found in the region at certain phases of the tide by Ingram[19,20] and Pingree and Griffith[21] may also be considered as another mechanism for intense mixing. In this area, the water at certain times is virtually homogeneous in a vertical section at 5°C and 28‰, even in the summer. Forrester[22] described elegantly the presence of strong internal tides which appear to be propagated from the head of the Laurentian trough. Both physical and biological observations show vertical oscillation and heights of tidal internal waves in this region up to 89 m.[18,23-25] Such vertical oscillation and intense mixing will bring a large flux of comparatively dense water into the surface layer. This cold and dense water contains a high concentration of nutrients and may explain the high level of biological productivity observed by Steven,[26] Sinclair et al.,[27] and DeLafontaine et al.[28] in the St. Lawrence estuary.

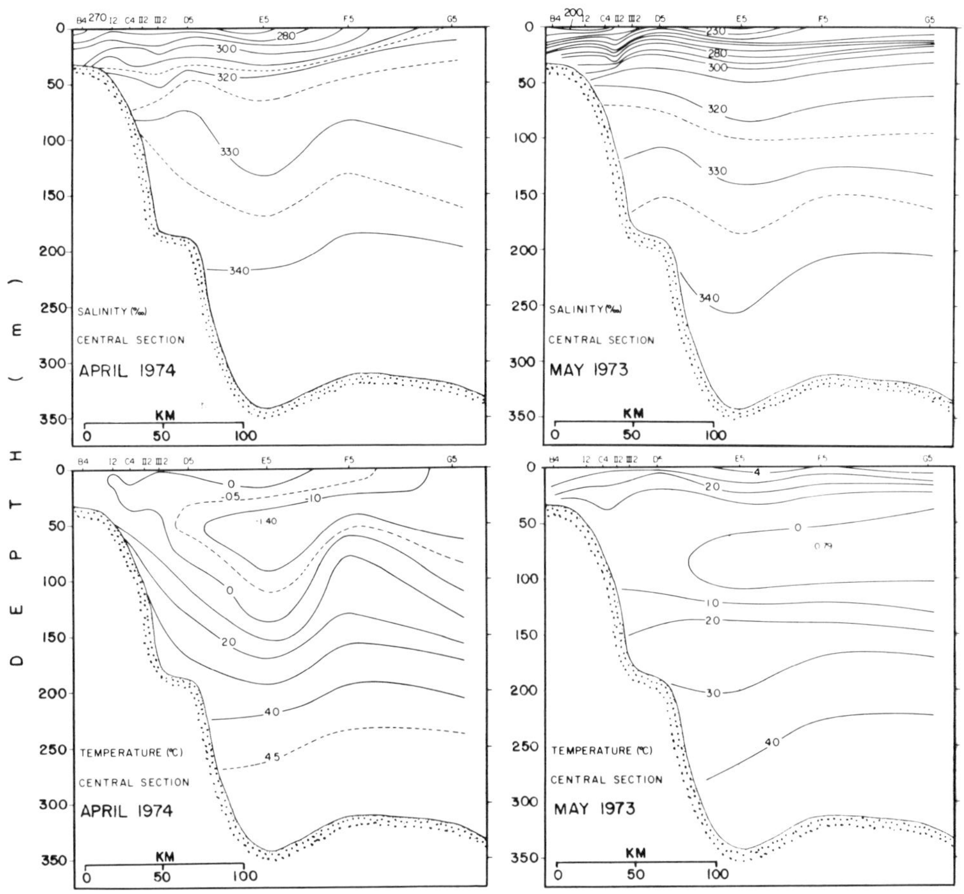

FIGURE 5. Vertical salinity and temperature distributions in the lower St. Lawrence estuary in May 1973 and April 1974.

The presence of internal oscillations with tidal periods,[29-30] and with periods less than semidiurnal tides,[23,31-33] have also been reported in other areas of the St. Lawrence estuary.

A recent detailed description of the general characteristics and seasonal variations of temperature, salinity, and density in the lower St. Lawrence estuary was given by El-Sabh.[1] Here we will concentrate on the transition period between April and May, during which maximum freshwater runoff occurs. Figure 5 shows longitudinal tidally averaged salinity and temperature distributions in the central section during these 2 months. Two layers are recognizable during late winter (April); a cold sub-zero upper layer with surface salinities increases seaward from 27 to 31.5‰, and overlies a deep warmer layer with temperatures less than 4.5°C. The more variable of these two layers is the upper one. Its characteristics are influenced strongly by land drainage and surface exchange with the atmosphere. In early spring (May), a strong thermocline is developed and, as a result of solar heating and increase in freshwater discharge, a warm water mass ~10 m thick, with temperatures on the order of 4°C, appears at the surface. This is underlain by cold water, with minimum temperature at a depth of about 75 m, which overlies the deep, warmer water. The input of more freshwater in May results in a decrease of surface salinity to values between 20 and 25‰ and an increase of the depth of the isohalines in the upper 100 m. The intermediate layer of the St. Lawrence estuary, which is formed by the advection of cold water from the Gulf of St. Lawrence during late winter,[34] has its coldest temperature of − 1.4°C in April, in the area of Rimouski.

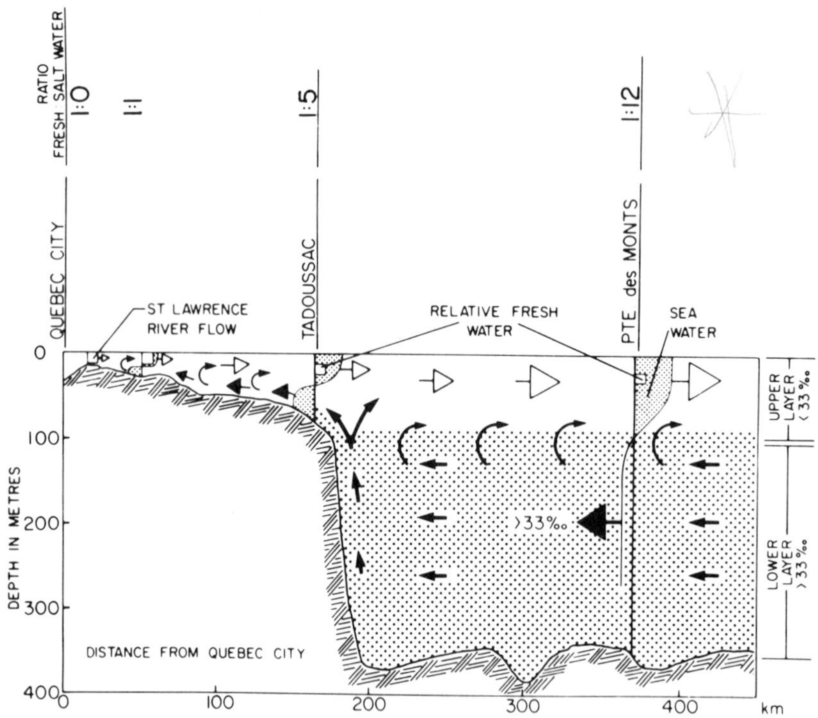

FIGURE 6. Schematic presentation of density current in the St. Lawrence estuary.[4]

Notice that fresher and warmer surface water in the lower estuary occurs at station E_5 in the vicinity of the Manicouagan Peninsula. This is due to the influence of the freshwater inflow from the B.O.M. north shore rivers. Furthermore, Figure 5 shows that the isohalines and isotherms in this layer, which were tilted upward at the head of the Laurentian trough in April, become almost horizontal in May. An annual signal of salinity and temperature at 50 m was also reported by El-Sabh.[1] Winter-spring salinity differences in the deep layer are insignificant, but the temperature decrases in May. El-Sabh[1] showed that the low frequency variability of temperature and salinity at 200 m depth is characterized by 2 to 4 month periods of fluctuation.

The salinity distributions along both shores of the lower estuary differ from that along the center. Usually, when looking down an estuary in the Northern Hemisphere, the less saline water tends to be held against the right side and more saline water held along the left side as a consequence of Coriolis acceleration. Neu,[8] El-Sabh[1,35] and Koutitonsky et al.[36] showed that the lower St. Lawrence estuary is far from ideal. It was shown on several occasions that in the area between Pointe-des-Monts and Rimouski, surface waters along the northern half are characterized by lower salinity, higher temperature, and therefore lower density compared to the south shore. This is clearly due to the freshwater inflow from the Saguenay and B.O.M. north shore rivers. The Gulf water, which has a surface temperature of 4.5°C and a salinity of more than 25‰ in May 1973, penetrates the estuary in mid-channel near the estuary mouth.

VI. ESTUARINE CIRCULATION

Under steady-state conditions, the general character of the tidally averaged estuarine circulation in the St. Lawrence is represented schematically in Figure 6, according to Neu.[8]

The vertical longitudinal circulation pattern shows a seaward-moving brackish upper layer and a landward-moving deep saline layer. In addition to this two-layer circulation, displacement during the tidal cycle creates strong cross-channel flows,[37] and a cyclonic circulation throughout the entire upper section of the estuary.[8] Recent studies, however, show that estuarine dynamics in the St. Lawrence are much more complex than previously considered.

From the examination of a complex array of current meters and temperature-salinity measurements made in the upper estuary between 1974 and 1977, Muir[38] concluded that 94% of the energy is contained within the tidal frequencies. Furthermore, the complex geometry of the upper estuary contributes to the existence of certain features such as the tidal elevation differences between north and south shores, deep penetration of isohalines along the north shore and different paths for tidal propagation during ebb and flood. According to Muir,[38] the two most important factors controlling circulation in the upper estuary are the tidally averaged density field and the internal wave field. The tidally averaged density structure is a tidal process because the exact amount and location of the mixing that occurs is controlled by the surface and internal tides. Most of the following information dealing with the dynamics of the upper estuary are taken from Muir.[38]

Surface tides which propagate up the Gulf of St. Lawrence and the lower estuary are unaffected by the density structure of the water, but are influenced a great deal by the topography of the channel and the Coriolis force. By the time the tides reach the head of the Laurentian trough, they already possess considerable complexity.[24,39] Internal tides are believed to be generated in this region by the interaction of surface tides and bottom topography.[22,40] When they arrive in the upper estuary, the tides are immediately split by Ile-aux-Lièvres. In the relatively shallow South Channel, considerable mixing occurs. Consequently, no internal tides can be propagated and the South Channel acts as an energy sink in the system and provides a source of well-mixed relatively warm and fresher water for the lower estuary. In the North Channel, both surface and internal tides are propagated without hindrance, since there is a well-developed vertical tidally averaged density structure. The energy distribution between the barotropic and the baroclinic waves, as well as the vertical and lateral modes which are present, will be determined by the conditions at the head of the Laurentian trough. Moreover, bottom topography of the upper estuary is such that internal tides should be reflected and may be in resonance with some of the topographic features. These resonances would not necessarily be present at all times, but when present could strongly affect the mixing and observed properties of the water at certain locations.

Fronts are a common feature in the area of Pointe-au-Pic on the changes of the tide. These are produced by the shears that are set up because of differences in time when the tide changes on the surface and at the bottom. The two-dimensional numerical model developed by Borne DeGrandpré and El-Sabh[41] and Borne DeGrandpré et al.[40] shows that this area is characterized by the divergence and convergence of currents which occur at times of slack water. The high-density gradients that are set up by the fronts could also have strong effects on the internal waves.

There are other nontidal processes which also have an effect on the mixing and on the tidally averaged density structure in the upper St. Lawrence estuary. The two most important of these are hydrological and meteorological conditions. Neu[8] has proposed that the intensity of the estuarine calculation is determined by the freshwater discharge. With geostrophic considerations in mind, this is a very reasonable conclusion, and for continuity to be maintained, there must be compensating currents either near the surface or at depth. On the other hand, meteorological conditions are variable throughout the upper estuary over any given time period. Aubin et al.[42] have discussed the effects of the wind on the surface currents in the Rivière-du-Loup area. In general, easterly winds prevail in the spring and westerly or southwesterly winds prevail during the summer, with mean wind speeds of about 8 m/sec. Due to the surface topography of the north side of the upper estuary, the winds blow more or less up and down the estuary, and the circulation and mixing effects of these winds must be considerable at times.

Seaward of the Saguenay River, the character and dynamics of the estuary change markedly. By utilizing all available current measurements and other information collected up to 1978, El-Sabh[1] proposed a sketch of a typical summer residual surface circulation pattern for the lower estuary. The outflow of freshwater along the south shore with speeds of 20 cm/sec, the existence of two gyres (a large anticyclonic one with its center between the Pointe-des-Monts and the Rimouski cross sections, and a smaller cyclonic eddy between Rimouski and Tadoussac), together with the northerly transverse current in the Ile-du-Bic region and the southerly transverse current at the estuary-gulf boundary near Pointe-des-Monts with speeds over 20 cm/sec, are the main features of this circulation pattern. The presence of the two gyres has recently been supported by current meter observations.[18,34,43,44] At mid-depth the mean flow decreases and adopts a cyclonic pattern while near the bottom the water flows upstream with speeds of 2 cm/sec and has some tendency toward the north shore.[44]

Based on the analysis of 4-month records of currents collected in 1979 in the area between Pointe-des-Monts and Mont-Joli, El-Sabh et al.[43] showed that most of the residual kinetic energy (E_r) is confined to the upper layer and decreases with depth. On the other hand, the tidal kinetic energy (E_t) is relatively uniform from surface to bottom. In general, E_r exceeds E_t in the upper layer, and is sometimes eight times greater, particularly during the spring when freshwater input reaches its maximum value. For high frequency fluctuations (less than 25 hr), E_t shows the dominant spectrum peak at the semidiurnal frequency band where most tidal energy is concentrated. Tidal energy for the axial component or current U is about 10 times greater than that for the cross-channel component (V). However, the energy in the subtidal frequency range shows the importance of axial current at the near-shore stations, while at the mid-estuary stations, the cross-channel component is more dominant. The energy density spectra of residual current increases toward the low frequency bands and becomes comparable in magnitude to the energy located at the semidiurnal frequency.

Energy can be transferred from the oscillatory tidal current into a residual motion, which is accomplished by the interaction of the tide with topography of the estuary. Murty and El-Sabh[45] numerically studied the tidally generated residual motion in the lower estuary and showed that it is about 4 to 5 cm/sec. While the oscillatory tidal currents do not show small-scale eddy patterns, the residual motion associated with tides consists of eddies with a scale of 12 to 30 km.

The space-time low frequency variability of currents in the lower St. Lawrence estuary was described by El-Sabh et al.[43] and Koutitonsky et al.[46] The results show the presence of 2 to 20 days fluctuations, superimposed on seasonal and longer time varying trends. Analysis of the 1979 current measurements by Koutitonsky et al.[46] revealed surface intensification, spatial coherence over 100 km, alongshore kinetic energy predominance in a broad frequency band (12 to 18 days) and varying offshore and vertical polarization. Dominant modal variance peakes were found at 8 days, with correlated variability at 3.5 days. Winds on opposite shores were found to be predominantly energetic in the same frequency band, oriented alongshore, and coherent with alongshore currents at 2.8 and 14 days.[43,46,47] Murty and El-Sabh[48] suggested that internal adjustment due to meteorological forcing is important in the lower estuary, and estimated that typical values for such adjustment would take 7 to 10 days, which is the frequency at which pressure systems travel over the estuary. Recently, it has been shown by several authors that meteorological forcing on a time scale of 2 to 20 days may have an important effect on estuarine dynamics (e.g., Elliott,[49] Smith,[50] Wang,[51,52] Weisberg,[53] and Wong and Garvine[54]).

Close examination of extensive satellite imageries, currents and CTD measurements taken in the lower St. Lawrence estuary in 1978 demonstrates the complexity of the circulation

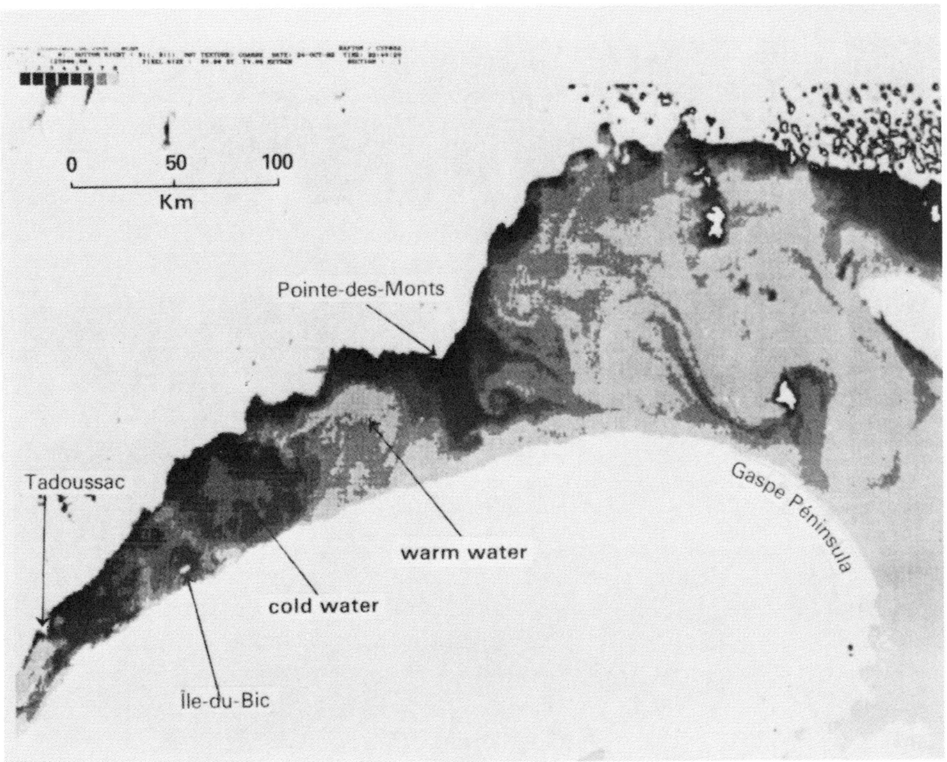

FIGURE 7. HCMM Satellite observations in the thermal infrared of the St. Lawrence estuary region on August 7, 1978. Lighter shades represent warmer temperatures.[61]

dynamics and reveals a number of features which have not been reported previously in the literature.[55] Within the time scale of 10 to 15 days, the estuary is subject to large perturbation and alternates between two modes of circulation pattern. The first mode (Figure 7), associated with periods of high sea level, is characterized by the presence of cold water mass at the head of the Laurentian trough, produced by the extensive mixing and internal tides described in Section VI; a series of cold and warm cores with scales of about 50 km; several lateral density fronts, including the Pointe-des-Monts cold front described by Tang,[56,57] and the coastal jet known as the Gaspé Current flowing seaward along the coast of the Gaspé Peninsula.[56,58-61] Furthermore, two other phenomena which have never been reported previously in the literature can be seen in the same satellite imagery. These are the warm triangular-shaped front(s) in the area of Ile-du-Bic and the two eddies on the Gulf side of the Pointe-des-Monts cold front with a space scale of about 15 km. The second mode (Figure 8), corresponding to periods when sea level is falling, features two distinct water masses separated by a longitudinal front with a wavelike shape at mid-estuary; the southern half of the lower estuary is colder than the northern part. The frontal demarcation line appears to have four crests and the horizontal distance between neighboring crests is about 75 km. This value corresponds to the wavelength of the topographic waves observed by Lie and El-Sabh.[62] The amplitude of this wavelike perturbation increases seaward with increasing estuary width. Similar structures were observed in the tank experiment of Stern et al.,[63] and are probably attributed to baroclinic instability.[64]

According to El-Sabh et al.,[43] surface circulation pattern in the very low frequency range

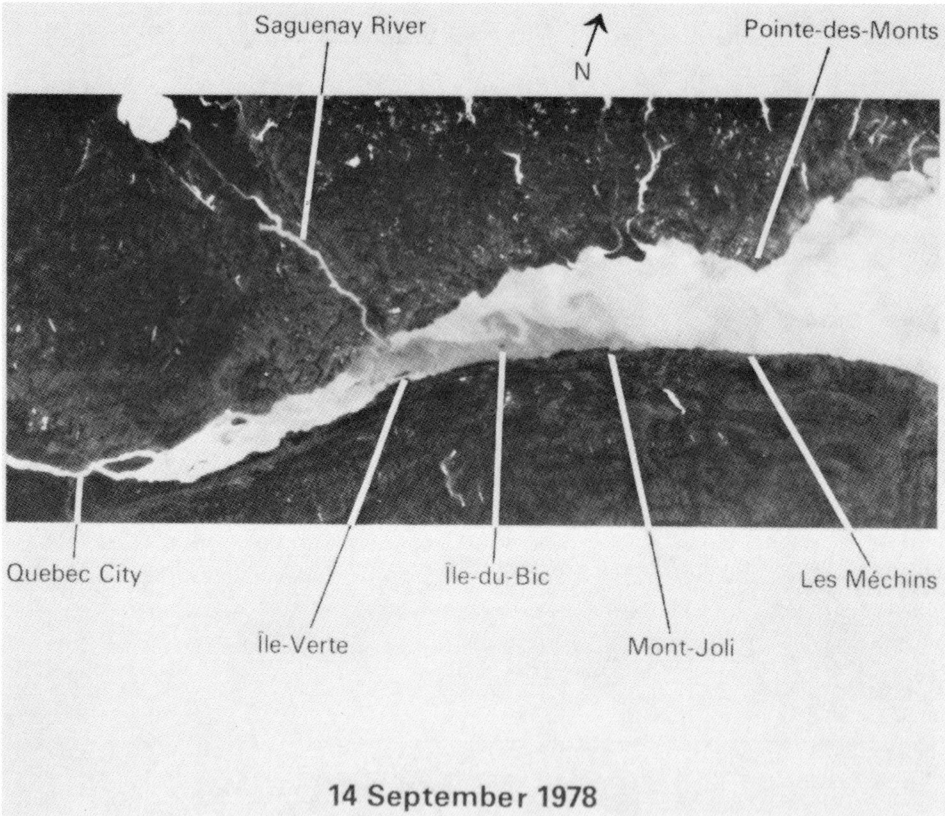

FIGURE 8. HCMM Satellite observations in the thermal infrared of the St. Lawrence estuary region on September 14, 1978. Lighter shades represent warmer temperatures.[61]

(70 to 80 days) is characterized by the presence of an estuary-wide gyre near the mouth (Figure 9); its intensity and rotation sense were found to be correlated to sea-level variations in the lower estuary. The mechanisms involved in the formation of the estuary-wide gyres and transverse currents have been a subject of speculation and several forces were proposed to explain their variability. Farquharson[65] was the first to point out the existence of an anticyclonic gyre, 50 km in diameter, from the path followed over a 5-day period by a parachute drogue set to drift at a depth of 75 m near the estuary mouth. He found that the intensity of the gyre is closely associated with the magnitude of the southerly transverse current observed in the Pointe-des-Monts section, which in turn occurs in response to the neap-spring tidal cycle. In a recent paper, Lie and El-Sabh[62] showed numerically that observed low frequency current variability in the lower estuary (Figure 9) could be explained by the superposition of two free baroclinic shelf waves traveling in opposite directions. Koutitonsky et al.,[46] on the other hand, showed that regulated freshwater, particularly from the Saguenay reservoirs, is discharged into the estuary as large pulses and not at a constant yearly rate. They suggest buoyancy forcing as another mechanism responsible for the observed very low frequency variability in velocity trends and residual gyre-like circulation near the mouth. Based on analysis of the 1982 current meter observations taken in the St. Lawrence estuary, Tee[66] also showed significant correlations between low frequency oscillation and freshwater pulses in the estuary.

The author believes, however, that the space-time variability of currents observed in the lower St. Lawrence estuary described in this paper is not the result of one or two forces, but rather is a result of the combined action of several factors such as the spring-neap tidal

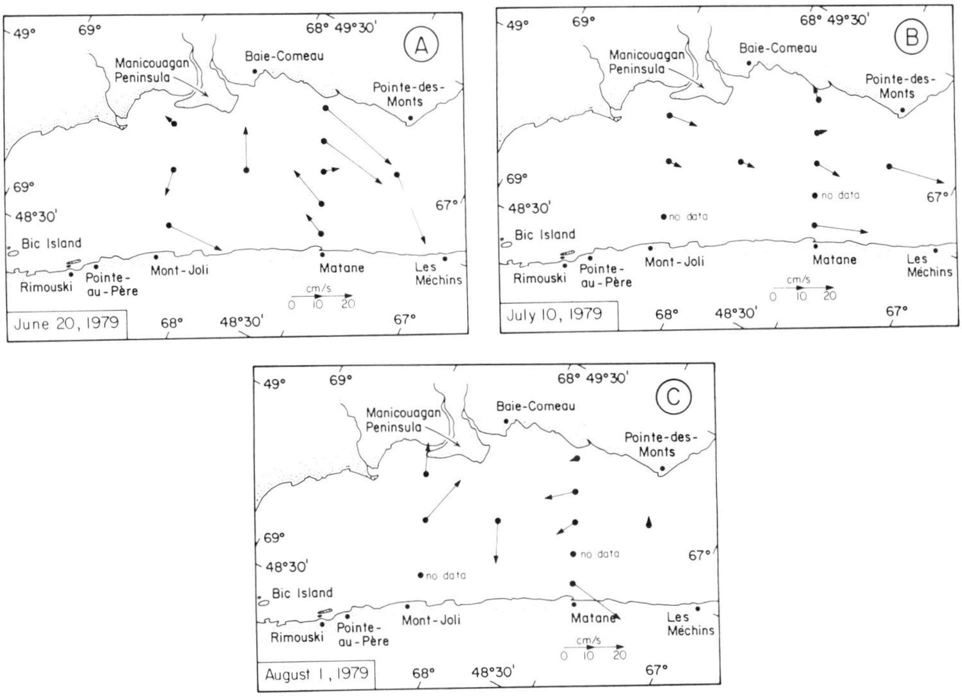

FIGURE 9. Modes of the low frequency circulation pattern in the lower St. Lawrence estuary in 1979.[43,62]

cycle, winds, atmospheric pressure gradients, freshwater discharge, strong mixing and upwelling processes at the head of the Laurentian trough, and topography. Broad scale and intensive field experiments are needed coupled with three-dimensional numerical models, to provide a quantitative assessment and better understanding of all the forces acting in the system.

ACKNOWLEDGMENTS

I wish to thank H. M. Edenborn and A. Mucci for their valuable comments on the manuscript, Laval Hotton for preparing the drawings, and Jocelyne Gagnon, for typing the manuscript. This work was supported by research grants to the author from the Natural Sciences and Engineering Research of Canada (No. A0073) and from the Foundation de l'Université du Québec à Rimouski.

REFERENCES

1. **El-Sabh, M. I.,** The lower St. Lawrence estuary as a physical oceanographic system, *Le Naturaliste Canadien,* 106, 55, 1979.
2. **Forrester, W. D. and Vandall, P. E., Jr.,** Ice volumes in the Gulf of St. Lawrence, Atlantic Oceanographic Laboratory, Bedford Institute, Report 68-7, 1968, 49.
3. **Bugden, G. L., Hargrave, B. T., Sinclair, M. M., Tang, C. L., Therriault, J.-C. and Yeats, P. A.,** Freshwater runoff effects in the marine environment: the Gulf of St. Lawrence example, Canadian Technical Report of Fisheries and Aquatic Sciences No. 1078, 1982, 89.
4. **Neu, H. J. A.,** Runoff regulation for hydropower and its effects on the ocean environment, *Can. J. Civil Eng.,* 2, 583, 1975.
5. **Neu, H. J. A.,** Man-made storage of water resources. A liability to the ocean environment? I and II, *Mar. Pollut. Bull.,* 13, 7, 44, 1982.

6. **Dickie, L. and Trites, L. M.,** The Gulf of St. Lawrence, in *Estuaries and Enclosed Seas,* Ketchum, B. H., Ed., Elsevier, Amsterdam, 1983, 403.
7. **Pritchard, D. W.,** What is an estuary: physical viewpoint, in *Estuaries,* Lauff, G. H., Ed., American Association for the Advancement of Sciences, Washington Publ. 83, 1967, 3.
8. **Neu, H. J. A.,** A study on mixing and circulation in the St. Lawrence estuary up to 1964, Atlantic Oceanographic Laboratory, Bedford Institute, Report 1970-9, 1970, 31.
9. **Hansen, D. V. and Rattray, J. M.,** New dimensions in estuary classification, *Limnol. Oceanogr.,* 11, 319, 1966.
10. **Meric, P.,** Circulation résiduelle dans l'estuaire moyen du Saint-Laurent et son influence sur les processus sédimentaires, Centre de recherches sur l'eau (CENTREAU), Université Laval, Québec. Report CRE-75-03, 1975, 68.
11. **Dyer, K. R.,** Lateral circulation effects in estuaries, in *Estuaries, Geophysics and the Environment,* Officer, C., Ed., National Academy of Science, Washington, D. C., 1976, 22.
12. **Larouche, P.,** Stratification et Équilibre Dynamique Latéral à L'embouchure de L'estuaire maritime du Saint-Laurent, M.Sci. thesis, *Université de Québec à Rimouski,* 1983, 84.
13. **Koutitonsky, V. G.,** Transport de masses d'eau à l'embouchure de l'estuaire du Saint-Laurent, *Le Naturaliste Canadien,* 106, 75, 1979.
14. **Levesque, L., Murty, T. S., and El-Sabh, M. I.,** Numerical modelling of tidal propagation in the St. Lawrence estuary, *Int. Hydrogr. Rev.,* 56, 117, 1979.
15. **Silverberg, N. and Sundby, B.,** Observations in the turbidity maximum of the St. Lawrence estuary, *Can. J. Earth Sci.,* 16, 939, 1979.
16. **Muir, L. R.,** Variability of temperature, salinity and tidally-averaged density in the middle estuary of the St. Lawrence, *Atmos. Ocean,* 19, 320, 1981.
17. **Pelletier, E. and Lebel, J.,** Hydrochemistry of dissolved inorganic carbon in the St. Lawrence estuary (Canada), *Estuarine Coastal Mar. Sci.,* 9, 785, 1979.
18. **Reid, S. J.,** Circulation and mixing in the St. Lawrence estuary near Ilet Rouge, Bedford Institute of Oceanography, Dartmouth, Report Series/BI-R-77, 36, 1977.
19. **Ingram, R. G.,** Characteristics of a tide-induced estuarine front, *J. Geophys. Res.,* 81, 1951, 1976.
20. **Ingram, R. G.,** Frontal characteristics at the head of the Laurentian Channel, *Le Naturaliste Canadien,* 112, 31, 1985.
21. **Pingree, R. D. and Griffiths, D. K.,** A numerical model of the M_2 tide in the Gulf of St. Lawrence, *Oceanologica Acta,* 3, 221, 1980.
22. **Forrester, W. D.,** Internal tides in the St. Lawrence estuary, *J. Mar. Res.,* 32, 55, 1974.
23. **Hassan, E. M.,** Coastal oceanography program in the St. Lawrence estuary in the spring-summer of 1973 and preliminary results of some aspects of the program, in *Proceedings of Workshop in the Physical Sciences in the Gulf and Estuary of St. Lawrence,* El-Sabh, M. I., Ed., Université du Québec à Rimouski, 1973, 2.
24. **Ingram, R. G.,** Influence of tidal-induced vertical mixing on primary productivity in the St. Lawrence estuary, *Mém. Soc. F. Sci. Liège,* 6, 59, 1975.
25. **Therriault. J.-C. and Lacroix, G.,** Nutrients, chlorophyll, and internal tides in the St. Lawrence estuary, *J. Fish. Res. Board Can.,* 33, 2747, 1976.
26. **Steven, D. M.,** Primary and secondary production in the Gulf of St. Lawrence, Report 26, Marine Science Center, McGill University, Montréal, 96, 1974.
27. **Sinclair, M., El-Sabh, M. I., and Brindle, J.-R.,** Seaward nutrient transport in the lower St. Lawrence estuary, *J. Fish. Res. Board Can.,* 33, 1271, 1976.
28. **DeLafontaine, Y., El-Sabh, M. I., Sinclair, M., Messieh, S., and Lambert, J.-D.,** Structure océanographique et distribution spatio-temporelle d'oeufs et de larves de poissons dans l'estuaire maritime et la partie ouest du golfe Saint-Laurent, *Sci. Tech. l'Eau,* 17, 43, 1984.
29. **Ingram, R. G.,** Internal wave observation off Ile Verte, *J. Mar. Res.,* 36, 715, 1978.
30. **Muir, L. R.,** Internal tides in the middle estuary of the St. Lawrence, *Le Naturaliste Canadien,* 106, 27, 1979.
31. **DeGuise, J. C.,** High Frequency Internal Waves in the St. Lawrence Estuary, M.Sci. thesis, McGill University, Montreal, Canada, 1977.
32. **Ingram, R. G.,** Vertical mixing at the head of the Laurentian Channel, *Estuarine Coastal Shelf Sci.,* 16, 33, 1983.
33. **Sinclair, M., Chanut, J.-P., and El-Sabh, M. I.,** Phytoplankton distributions observed during a 3 1/2 days fixed station in the lower St. Lawrence estuary, *Hydrobiologia,* 75, 129, 1980.
34. **Ingram, R. G.,** Water mass modification in the St. Lawrence estuary, *Le Naturaliste Canadien,* 106, 45, 1979.
35. **El-Sabh, M. I.,** Circulation pattern and water characteristics in the lower St. Lawrence estuary, in *Proc. Symp. Modelling of Transport Mechanisms in Oceans and Lakes,* Murty, T. S., Ed., Burlington, Ontario, 1977, 43, 243.

36. **Koutitonsky, V. G., Cossa, D., Poulet, S., El-Sabh, M. I., Piuze, J., and Chanut, J. P.,** Flux de matières particulaires dissoutes et de traces métalliques à l'embouchure de l'estuaire du Saint-Laurent, Ministère des Approvisionnements et Services, Contrat ISD-78-00066, *INRS-Océanologie, Université du Québec à Rimouski,* 1980, 253.
37. **D'Anglejan, B. F. and Ingram, R. G.,** Time-depth variations in the tidal flux of suspended matter in the St. Lawrence estuary, *Estuarine Coastal Mar. Sci.,* 4, 401, 1976.
38. **Muir, L. R.,** Internal tides in a partially mixed estuary, *Ocean Science and Surveys, Central Region,* Report No. 9, Canada Centre for Inland Waters, Burlington, Ontario, 177, 1982.
39. **Godin, G.,** La marée dans le golfe et l'estuaire du Saint-Laurent, *Le Naturaliste Canadien,* 106, 105, 1979.
40. **Borne DeGrandpré, C. de, El-Sabh, M. I., and Salomon, J. C.,** A two-dimensional numerical model of the vertical circulation of tides in the St. Lawrence estuary, *Estuarine Coastal Shelf Sci.,* 12, 375, 1981.
41. **Borne DeGrandpré, C. de and El-Sabh, M. I.,** Etude de la circulation verticale dans l'estuaire du Saint-Laurent au moyen de la modélisation mathématique, *Atmos. Ocean,* 18, 304, 1980.
42. **Aubin, F., El-Sabh, M. I., and Murty, T. S.,** Numerical simulation of the movement and dispersion of oil slicks in the upper St. Lawrence estuary: preliminary results, *Le Naturaliste Canadien,* 106, 37, 1979.
43. **El-Sabh, M. I., Lie, H.-J., and Koutitonsky, V. G.,** Variability of the near-surface residual current in the lower St. Lawrence estuary, *J. Geophys. Res.,* 87, 9589, 1982.
44. **Koutitonsky, V. G. and El-Sabh, M. I.,** Estuarine mean flow estimation revisited: application to the St. Lawrence estuary, *J. Mar. Res.,* 43, 1, 1985.
45. **Murty, T. S. and El-Sabh, M. I.,** Tidally-generated residual motion in the St. Lawrence estuary, in *Proceedings 13th Annual Simulation Symposium,* Boyd, V. P., Cummings, R., Hammer, C., and Malamphy, W., Eds., 1980, 127.
46. **Koutitonsky, V. G., Wilson, R. E., and El-Sabh, M. I.,** On low frequency current variability in the lower St. Lawrence estuary, Summer 1979, presented at 19th Annual Congress, Canadian Meteorological and Oceanographic Society, Montreal, June 12—14, 1985, 51.
47. **El-Sabh, M. I. and Gagnon, M.,** Tidal and low-frequency variability in the lower St. Lawrence estuary, *Estuaries,* 8, 2B, 68A, 1985.
48. **Murty, T. S. and El-Sabh, M. I.,** Transverse currents in the St. Lawrence estuary: a theoretical treatment, in *Transport Processes in Lakes and Oceans,* Gibbs, R. G., Ed., Plenum Press, New York, 35, 1977.
49. **Elliott, A.,** Observations of the meteorologically induced circulation in the Potomac estuary, *Estuarine Coastal Mar. Sci.,* 6, 285, 1978.
50. **Smith, N. P.,** Meteorological and tidal exchanges between Corpus Christi Bay, Texas, and the Northwestern Gulf of Mexico, *Estuarine Coastal Mar. Sci.,* 5, 511, 1977.
51. **Wang, D. P.,** Subtidal sea level variations in the Chesapeake Bay and relations to atmospheric forcing, *J. Phys. Oceanogr.,* 9, 413, 1979.
52. **Wang, D. P.,** Wind-driven circulation in the Chesapeake Bay, winter 1975, *J. Phys. Oceanogr.,* 9, 564, 1979.
53. **Weisberg, R. H.,** The nontidal flow in the Providence river of Narragansett Bay: a stochastic approach to estuarine circulation, *J. Phys. Oceanogr.,* 6, 721, 1976.
54. **Wong, K.-C. and Garvine, R. W.,** Observations of wind-induced, subtidal variability in the Delaware estuary, *J. Geophys. Res.,* 89, 10589, 1984.
55. **El-Sabh, M. I.,** Variability of surface currents in a wide and deep estuary: the St. Lawrence as a case study, *Estuaries,* 8, 28, 67A, 1985.
56. **Tang, C. L.,** Mixing and circulation in the northwestern Gulf of St. Lawrence: a study of a buoyancy-driven current system, *J. Geophys. Res.,* 85, 2787, 1980.
57. **Tang, C. L.,** Cross-front mixing and frontal upwelling in a controlled quasi-permanent density front in the Gulf of St. Lawrence, *J. Phys. Oceanogr.,* 13, 1468, 1983.
58. **Benoit, J., El-Sabh, M. I., and Tang, C. L.,** Structure and seasonal variations of the Gaspé Current, *J. Geophys. Res.,* 90, 3225, 1985.
59. **El-Sabh, M. I.,** Surface circulation pattern in the Gulf of St. Lawrence, *J. Fish. Res. Board Can.,* 33, 124, 1976.
60. **El-Sabh, M. I. and Benoit, J.,** Variabilité spatio-temporelle du courant de Gaspé, *Sci. Tech. l'Eau,* 17, 55, 1984.
61. **Lavoie, A., Bonn, F., Dubois, J. M., and El-Sabh, M. I.,** Structure thermique et variabilité du courant de surface de l'estuaire maritime du Saint-Laurent à l'aide d'images du Satellite HCMM, *Can. J. Remote Sensing,* 11, 70, 1985.
62. **Lie, H.-J. and El-Sabh, M. I.,** Formation of eddies and transverse currents in a two-layer channel of variable bottom with application to the lower St. Lawrence estuary, *J. Phys. Oceanogr.,* 13, 1063, 1983.
63. **Stern, M. E., Whitehead, J. A., and Hua, B.-L.,** The intrusion of a density current along the coast of a rotating fluid, *J. Fluid Mech.,* 123, 237, 1982.

64. **Griffiths, R. W. and Linden, P. F.,** The stability of buoyancy-driven coastal currents, *Dyn. Atmos. Oceans,* 5, 281, 1981.
65. **Farquharson, W. I.,** St. Lawrence estuary current surveys, Bedford Institute of Oceanography, Report Series 66-6, 84, 1966.
66. **Tee, K.,** The 1982 Current Meter Observations in the St. Lawrence Estuary, Presented at Second Symposium on the Oceanography of the St. Lawrence Estuary, Université Laval, Quebec, May 5—6, 1984.

Chapter 16

OCEANOGRAPHIC CHARACTERISTICS OF THE SEINE ESTUARY

Jean Claude Salomon

TABLE OF CONTENTS

I. INTRODUCTION

The river Seine is located in the northern part of France (Figure 1). With a hydrographic basin of 74,250 km^2, and an average flow of 450 m^3/sec, it takes its place among the medium rivers. However, as it flows through the city of Paris and drains a heavily urbanized and industrial basin (30% of the country's population, and 40% of its economic activity), its role is much more important than the preceding numbers could mean. Two other important cities are situated along the lower course of the river, i.e., along the estuary: Le Havre at the entrance and Rouen 130 km upstream.

In the past, man has used the river for his own activities and has attempted to modify it. As a consequence, many *in situ* measurements and theoretical studies have been conducted by port authorities, engineering companies, and universities. Bathymetric maps, for example, have been available since the 17th century, allowing for compilations of the historical evolution of the estuary. This favorable situation provides us with the data for reviewing quite easily its main hydraulic and sedimentological characteristics.

II. GEOGRAPHY

The Seine estuary opens into the English Channel at the eastern part of the Bay of Seine (Figure 1). As salinity values lower than regional marine characteristics are observable in the eastern part of the bay, the seaward limit of the estuary may be defined some 20 km from Le Havre. At the other extreme the fluvial limit is easy to determine: a dam stops the tide propagation at Poses, 163 km upstream of Le Havre. Thus limited, the estuary is 180 km long for a width of 40 km at the seaward limit, but only a few hundred meters in most of its length.

There are three distinct portions of the estuary:

1. Seaward of Le Havre's meridian, the estuary is wide, with water depths increasing from 5 to 20 m (respectively, to the lowest low water level) in the ancient Seine valley.
2. Upstream of Honfleur, the internal estuary is narrow and restricted to a navigation channel with a constant depth of 6 m.
3. The transition zone between the two domains, from Le Havre to Honfleur, where the channel is progressively separated from adjacent shoals by two jetties, over a distance of 15 km or so.

The present shape of the estuary is entirely man-made for navigational purposes. Figure 1 reveals how different the natural estuary was 150 years ago: the center part consisted of shallow meandering channels surrounded by extensive tidal flats.

III. RIVER INPUT

Minimum river flow occurs in the summer, from May to October, with monthly mean values of about 250 m^3/sec. The annual average is actually 450 m^3/sec and seems to have decreased lately. Winter is the period of highest river discharge, with a maximum in December and January, which usually exceeds 1200 m^3/sec. The greatest known flood is 2500 m^3/sec.

IV. TIDE AND CURRENTS

The most noticeable characteristics of this estuary are its tidal wave peculiarities. The Atlantic tide, essentially semidiurnal, enters the English Channel at its west opening and

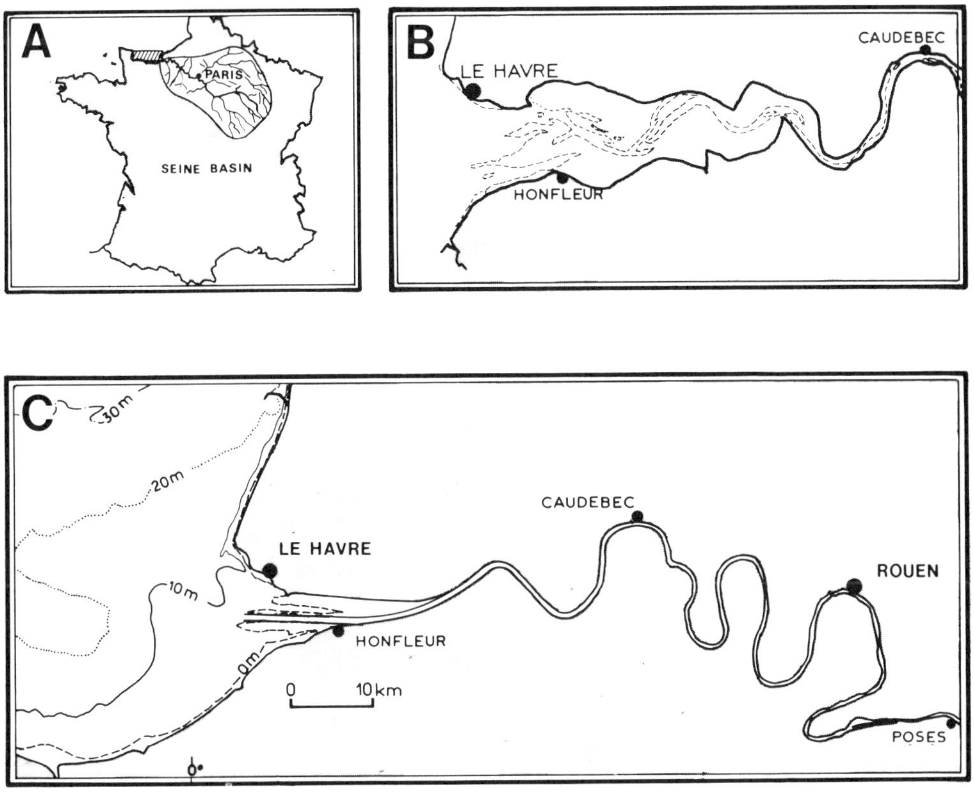

FIGURE 1. Location maps: (A) hydrographic basin and bay of Seine, (B) the central part of the estuary in 1834, and (C) the present estuary.

propagates toward the North Sea. As the Channel rapidly narrows, reflection becomes important, and the resulting system mainly corresponds to a Kelvin amphidromy. The wave, sinusoidal when leaving the ocean, becomes rapidly distorted because of small depth propagation and partial reflections.

The observed tidal shape near Le Havre is shown in Figure 2: water level rises very suddenly, then flattens and remains constant for 2 hr or 2 hr, 30 min. During spring tide it can even show maxima separated by a small intermediary minima. This phenomenom is mathematically interpreted by the apparition of harmonics M_4, M_6, M_8, and MS_4. The mean range is 5 m, and reaches 8.5 m during spring tides.

Farther upstream in the estuary, the tidal curve continues distorting (Figure 2). The beginning of the ascending phase becomes shorter and, in some circumstances of great amplitude, gives rise to a bore in the region of Rouen.

To facilitate navigation to the port of Rouen, and upstream to the port, man has intensively modified the estuarine geometry. Severe problems occurred during neap tide because of decreased tidal range. The estuary has thus been modified in order to maintain, and better amplify, the tidal range during its propagation. The modifications result in steadily decreasing cross sections in the estuary which compensate for the loss of energy due to friction.

The estuary is now almost synchronous during neap tide, whereas during spring tide, because of the high increase of energy losses proportional to square velocities, it is hyposynchronous (Figure 3).

The dam, located at Poses, also contributes to maintaining large tidal amplitudes in the upper reaches: it completely reflects the incoming wave and creates a partially standing system, amplifying its range immediately downward.

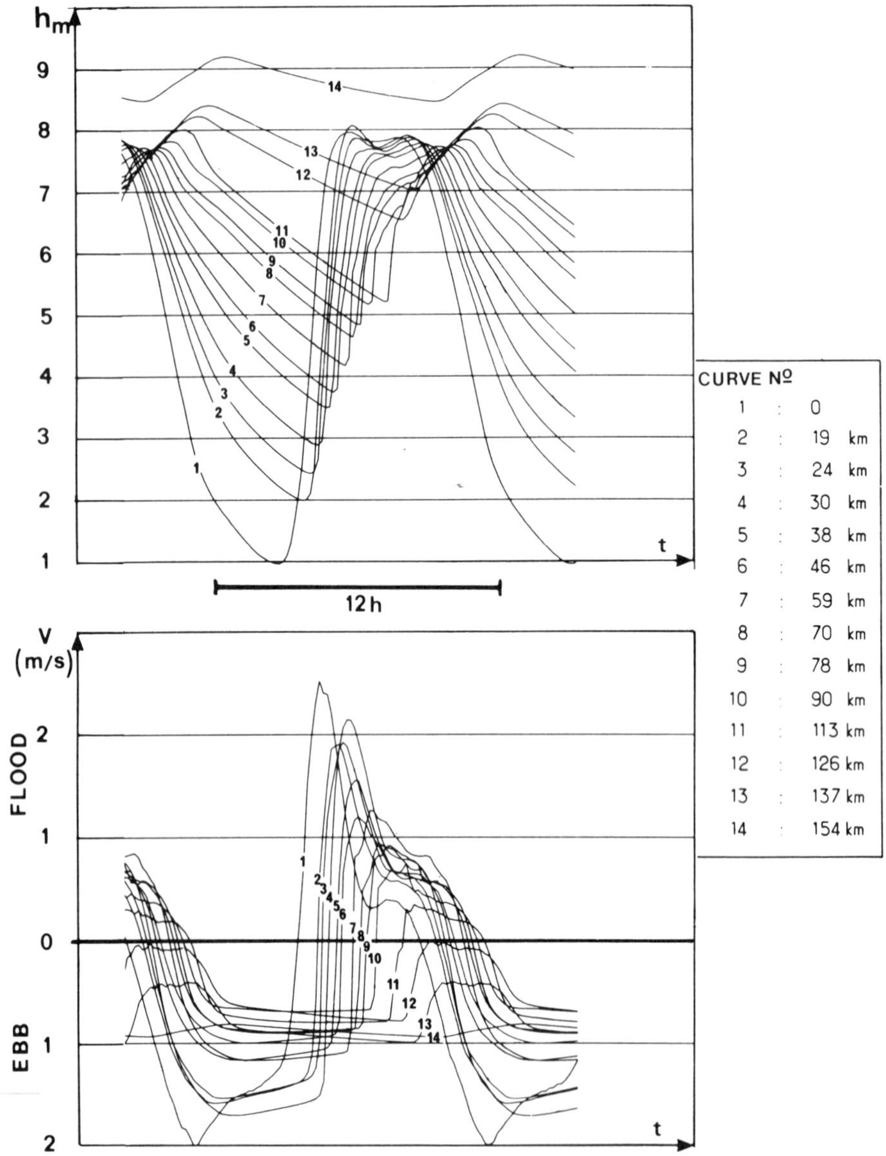

FIGURE 2. Tide levels and velocities along the estuary (spring tide and high river flow; distances refer to Honfleur).

As is generally the case in estuaries, in the upper reaches low water levels during neap tides are lower than during spring ones, and the contrary near the mouth. The transition point is situated in a rather unusual position in the Seine Estuary: very close to the sea.

Maximum intensity of currents occurs during flood, except in the navigation channel during periods of high river flow and/or weak tides (Figures 2 and 4). In the outer estuary, as in the bay, the early flood current, which corresponds to the rapid rise of levels, is directed east, towards the inside of the estuary. Because of the funnel shape of the estuary, the tidal prism is fairly low (140 millions of m^3 upstream of Honfleur), and the mouth is wide. It is thus quite rapidly filled, and then flood turns north, driving waters out of the bay. This secondary flow is locally known as the ''Verhaule'' current. Ebb current, oriented west, is more regular in intensity and direction. The resultant hodograph rotates in the positive

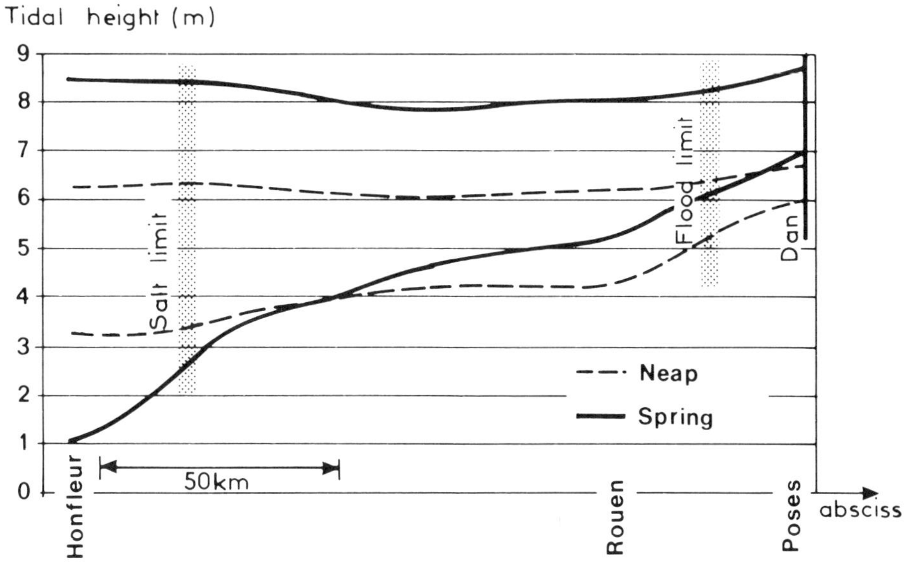

FIGURE 3. High and low tide levels vs. abscissa.

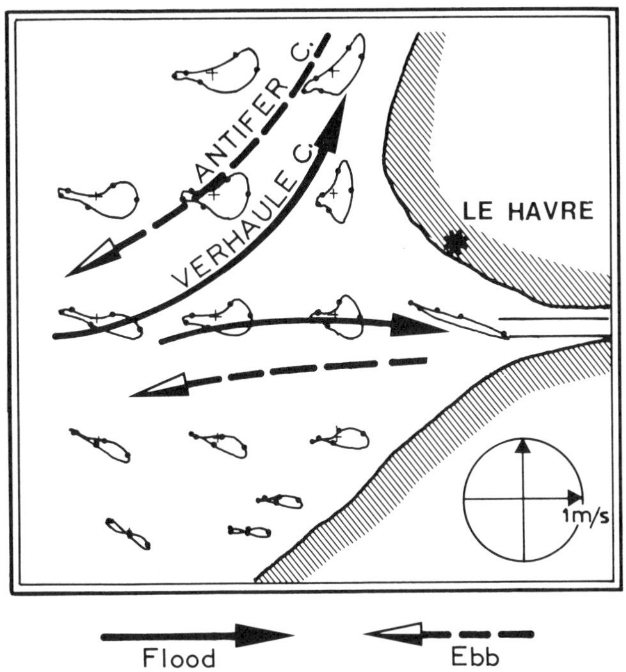

FIGURE 4. Currents in the external estuary.

direction, and generally exhibits three maxima (two for flood and one for ebb). Further upstream, the peak of flood gradually diminishes, and finally, under normal fluvial and marine conditions, flood current completely disappears some 25 km downstream of Poses.

V. SALINITY INTRUSION AND WATER STRATIFICATION

Due to the above-mentioned modifications that restrict the inner estuary to a navigation channel, the volume of water penetrating from the sea has been severely reduced, and so has the intrusion of salt. At present, the salinity intrusion extends 40 to 50 km upstream of Le Havre during low river-flow periods, and only 10 to 15 km during flood conditions (Figure 5). Toward the sea (as was already noted), low salinity values are observable all over the eastern part of the bay.

Vertical salinity gradients are quite low. Stratification indexes such as the ratio between tidal prism and river inflow, or the rate of energy dissipated by friction vs. the change of potential energy through mixing, class the Seine either among the well-mixed estuaries, or in the upper limit of partly mixed ones, depending on fluvio-tidal conditions.[1] This classification is confirmed by *in situ* measurements, except in a small area just seaward of the jetties, where a bathymetric step induces a localized zone of higher salinity stratification.

Correspondingly, instantaneous current shears are restricted to the vicinity of the bottom,[1] and velocity profiles essentially obey either the well-known logarithmic law:

$$U = (V^*/K) \text{ Log } (Z/XZo), \text{ with } Zo = 1 \text{ cm}$$

or an exponential one:

$$U = aZ^{**}b, \text{ with } b = 0.25$$

In the outer estuary a slight veering of the vertical profiles is due to lesser intertial forces near the bottom.[2]

The residence time of water and dissolved material has been calculated by mathematical models: from Poses to Le Havre it is about 10 days under normal conditions. This fairly low value corresponds to the limited volume of water contained in the inner estuary.

VI. RESIDUAL VELOCITIES

A macrotidal estuary such as the Seine with a highly distorted tidal curve illustrates the poor information one can obtain from Eulerian residual velocities. Flood currents occur mainly during high tide, while the water level is significantly higher than it is during ebb (Figure 2). The difference here can be as great as 2 m, thus causing a net predominance of Eulerian velocities directed toward the sea, even if the river discharge was insignificant (Figure 5). These Eulerian velocities, always directed seaward in the channel, are by no way representative of the real water displacement.

This characteristic is still amplified in the outer estuary, as variations in the velocity direction cause large horizontal gradients. It particularly occurs during the flood tide when waters enter both the inner estuary, and move northward toward the eastern English Channel, or during the ebb in the region where "Antifer" current joins the ebbing estuarine water. For the same reason the position of a particle in a water column with respect to time is of utmost importance for specifying its trajectory; which complicates remarkably the interpretation of the Langrangian drift. In a schematic way let us say that freshwater essentially goes northwest, and that a main saltwater flux enters the outer estuary from the north, along the right side.

This long-term circulation is affected by the fortnightly succession of neap and spring tides. During increasing tide ranges the mean tidal level rises in the estuary, with a corresponding net storage of water equivalent to a discharge of 100 m^3/sec or so, and inversely during the transition between neap to spring. Such a flow can be of the same order as the river discharge during summer, thus modifying significantly the residual circulation.[3] Presumably wind stress also strongly affects water circulation in the external estuary but, at present, very few details are available on the effects of wind.

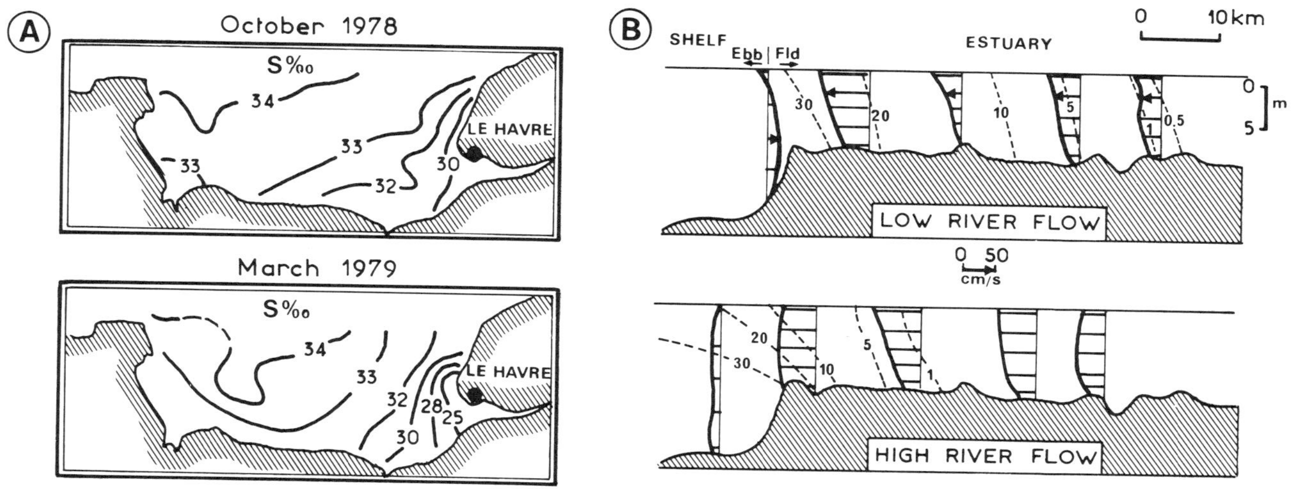

FIGURE 5. Salinity distribution: (A) external estuary and bay and (B) internal estuary; with residual velocities.

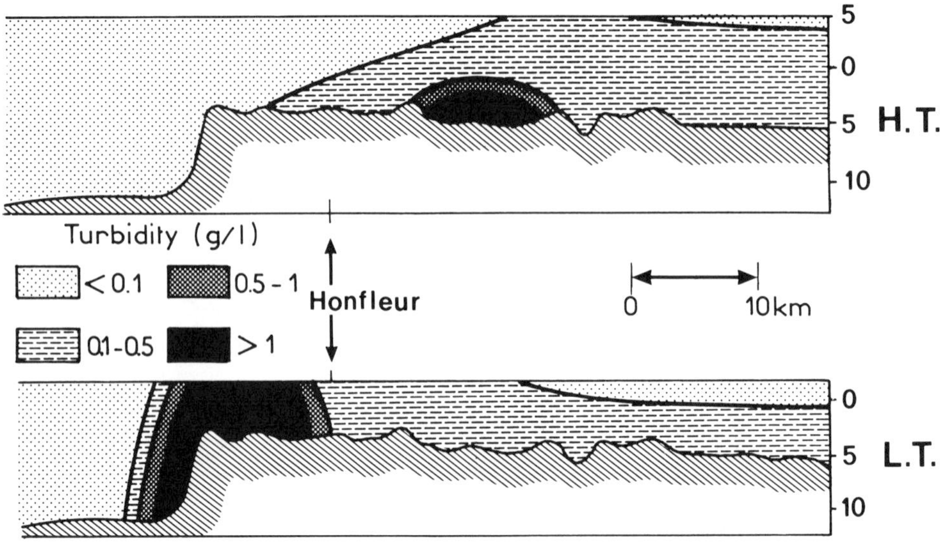

FIGURE 6. Turbidity maxima (spring tide and high river flow).[4]

VII. SEDIMENT DISTRIBUTION AND MOVEMENT

Sediments in the mouth consist of fine sands with localized patches of gravel. Mud zones are also present located on either side of the channel jetties. These mud patches sometimes enter Le Havre harbor which, of course, causes severe dredging problems. However, the accumulation of sands in the harbor is the most significant and routine cause for maintenance dredging.

Fine marine sand (D = 0.15 mm) is present in great quantities in the eastern bay. As flood velocities are higher than ebb velocities everywhere in the bay, sands predominantly penetrate the estuary. The sharp peak in flood velocity shown Figure 2, is very efficient on that point and, depending on the ratio between tide amplitude and river flow, moves sand a few tens of kilometers upstream of Le Havre. This is the basic reason why, in the past, this region has constantly been changing, encumbered with moving sand banks migrating upstream during low flows, and flushed away in winter.

Jettying the channel has greatly reduced the shoaling problems by decreasing velocities on either side of it, and correspondingly stabilizing sand banks in these areas. As a result deposition in the main channel may have increased because the diminishing channel width would have amplified velocities, therefore increasing the sediment transport directed into the estuary. A special design of these jetties based on the local tide peculiarities has prevented this from occurring.

Engineers have used the large asymmetry of the tidal curve, and especially the persistence of high tide, to aid dredging of the channel. Building weirs emerged at high tide, but submerged at low tide, allows the possibility for flood waters to pass over the weirs, but maintains ebb in the channel, between the jetties. Consequently, hydraulic sections are greater during flood than ebb, and flood velocities reduced compared to ebb velocities. This creates a natural way of sweeping the channel of undesirable sands in the central region of the estuary, thus preventing them from penetrating further upstream.

Now, maintenance dredging problems are restricted to the entrance of the jetties, where sand coming from the bay is deposited and accumulates.

A turbidity maximum exists in the estuary, in the jettied region and upstream of it (Figure 6), where tidal velocities are maxima.

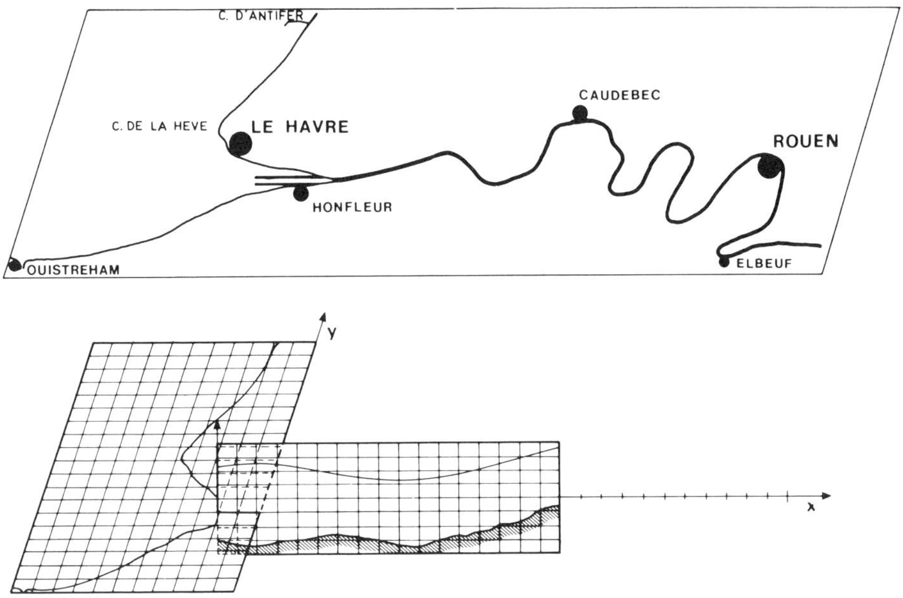

FIGURE 7. Schematization of the Seine estuary by a set of three numerical models.

The amount of sediment in suspension is of the order of 40,000 tons during neap tides and high river discharge,[4] and only 2,000 tons during neap tides. The daily fluvial influx ranges between 200 tons to 300 tons during low river flow, and 10,000 tons during highest river flow. The average residence time for these sediments may be very short.

Recent studies show that the reason sediment deposition persists inside the estuary while the residual velocities are directed seaward, involves instantaneous dynamical processes:[5,6]

The asymmetry of tidal currents causes more erosion during flood than during ebb, and combines with the different duration of slack waters at high and low tide to allow more sediment to be deposited and consolidated in the upstream limit of their tidal excursion than in the lower part of it. The varying amplitude of velocities along the abscissa and their vertical orientation directed upward during the transition from ebb to flood whereas in the lower part of the estuary it is directed downward after flood are other efficient mechanisms contributing to trap sediments inside the estuary.

All of these instantaneous dynamical effects that have no relation with the classical density circulation, help to explain the relative persistence of fine sediments in the estuary although the residual velocity of water is oriented toward the sea.

VIII. MODELS

Most estuarine modifications that were done in the past to facilitate industrial use of the waterway were previously tested through physical modeling. The present estuarine morphology, which creates a natural mechanism of auto dredging the lower part of the channel while maintaining a large tidal amplitude at Rouen although the natural width of the estuary is very narrow, results from the use of this technique.

Recently, new problems and developments related to water quality, and environment in a more general sense, have led to interdisciplinary studies and the elaboration of a complete set of mathematical models.

In agreement with estuarine morphology, the models are unidimensional in the upper reaches, bi-dimensional in a vertical plane in the central part, and bidimensional in a horizontal plane in the outer part of the mouth (Figure 7). The mechanisms involved in the models are hydrodynamics, dispersion of dissolved substances, sediment movements, and water quality.[2-7] Even more recent is the development of a three-dimensional model of the bay,[8] although it has not yet reached a sufficient predictive state to be used as a management tool.

IX. CONCLUSIONS

The Seine is a good example of macrotidal estuary with tidal range and depths of the same order of magnitude. The amplitude and specific shape of the tidal curve combine with relatively low river inputs to give the estuary essentially marine dynamic characteristics. However, through artificial modifications that reduce the estuary to a simple channel, man has restricted the intrusion of saltwater from the marine environment, thus geographically separating dynamic and densimetric processes. As was concluded by recent studies,[3] "Man has changed the Seine from a relatively open system with saltwater intruding landward, to a narrow river mouth debouching freshwater and fluvial sediments directly into the sea".

This new estuary is far from being in an equilibrium state, and sediments escaping from the channel are presently accumulating in the lateral parts of the mouth. The result is the beginning constitution of a new estuary in the eastern bay. A very interesting perspective will be to see how nature and man's actions are going to combine and mold this new Seine estuary.

REFERENCES

1. **Salomon, J. C.,** Modèle Mathematique de la Propagation de la Marée en Estuaire et des Transports Sableux Associés: Application aux estuaries de la Loire et de la Seine, Thèse Dictorat, University of Brest, France, 1976.
2. **Salomon, J. C. and Le Hir, P.,** Etude de l'estuaire de la Seine. Modélisation numérique des phénomènes physiques, *Univ. Brest,* 286, 1981.
3. **Avoine, J., Allen, G. P., Nichols, M., Salomon, J. C., and Larsonneur, C.,** Suspended sediment transport in the Seine estuary, France: effect of man made modifications on estuary-shelf sedimentology, *Mar. Geol.,* 40, 119, 1981.
4. **Avoine, J.,** Etude hydrosédimentaire de l'estuaire de la Seine, Thèse University of Caen, France, 1980.
5. **Allen, G. P., Salomon, J. C., Bassoullet, P., de Penhoat, Y., and de Grandpre, C.,** Effects of tides on mixing and suspended sediment transport in macrotidal estuaries, *Sediment. Geol.,* 26, 69, 1980.
6. **Salomon, J. C.,** Modelling turbidity maximum in the Seine estuary, in *Echohydrodynamics,* Elsevier, Amsterdam, 1981.
7. **Romana, L. A.,** Un modèle mathématique de nitrification en basse Seine, *Symp. 17ème Journées de l'Hydraulique,* Nantes, 5, 6, 1982.
8. **Thouvenin, B. and Salomon, J. C.,** Modèle tridimensionnel de circulation et de dispersion en zone côtière à marée. Premiers essais: cas schématique et Baie de Seine, *Océanol. Acta,* 7, 4, 417, 1984.

Chapter 17

CONSEQUENCES OF DREDGING

Maynard M. Nichols

TABLE OF CONTENTS

I. INTRODUCTION

The ever increasing demand for creating new channels for shipping, boating, and naval defense has made it necessary to gain a more complete understanding of the consequences of dredging on estuarine hydrodynamics and sedimentation. Dredging deepens channels and increases the estuary volume. The bathymetry is smoothed as bars and shoals are removed and as channels are straightened. Dredging, in turn, requires disposal of the enormous amounts of material removed. Commonly, this material is placed alongside the channel on bordering shoals, or on marshes and behind dikes to reclaim intertidal areas. What, then, are the long-term consequences of deepening, smoothing, and narrowing an estuary on the circulation and salinity? How do the hydrodynamic changes affect sediment accumulation rates and patterns? It is essential to understand and to be able to predict these consequences if we are to manage estuaries wisely.

This paper shows the effects of dredging on hydrodynamic variables through a review of concepts and case studies. Most of the information comes from engineering studies in which proposed physical alterations resulting from dredging are fitted into hydraulic or numerical models. After natural characteristics of the tide, current, salinity, or sedimentation patterns are reproduced and verified in the models, tests are conducted over a range of river inflow and tidal amplitude to detect differences before and after the proposed alteration. Although the models are constructed and verified from field observations, very few test results have been verified by subsequent field observations after dredging is performed.

Two aspects of estuarine hydrodynamics are affected by changes in geometry caused by dredging: (1) the tidal hydraulics including landward propagation of ocean tides, tidal discharge, and channel stability, and (2) the residual density flow or estuarine circulation.

II. TIDAL EFFECTS

A. Landward Modifications

When ocean tides are propagated into a shallow estuary, they become modified by three different processes

1. Frictional damping on the bottom
2. Landward constriction by shoaling or convergence in the channel
3. Reflection from shoals or from the estuary head[1]

Friction with channel boundaries dissipates the energy of the tide wave and the tidal amplitude decreases exponentially landward (Figure 1A) as a function of the relation:

$$A = A_o e^{-kx} \tag{1}$$

where A is the amplitude at x distance from the mouth; A_o is the amplitude at the mouth; and k is a friction coefficient, which is a function of the tidal current velocity and frictional resistance indexed by the Chezy coefficient.[2] As amplitude decreases landward and current velocity diminishes, fine sediment is likely to be retained and deposited in landward reaches. When a long tidal channel is enlarged or straightened by dredging, frictional dissipation is reduced. Consequently, the tidal amplitude increases in upstream reaches (Figure 1C). Tides can propagate faster at all stages of the tide, the effect being greatest near low water.[3]

After enlargement of the Delaware River near Philadelphia following deepening from 3.6 to 6.5 m and removal of several islands, the tidal range at the estuary head increased and the mean low water datum was depressed by about 30 cm at Trenton.[4] The effect of dredging on increasing the tide range is confirmed by numerical model simulations.[5] The model

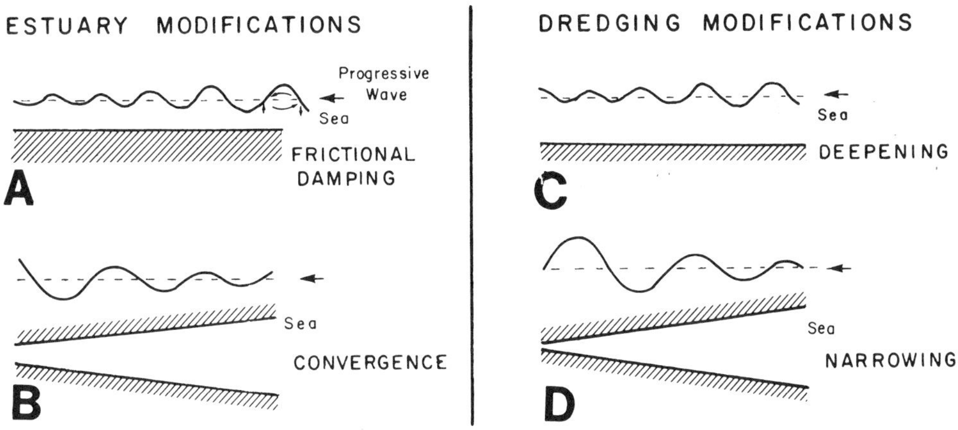

FIGURE 1. Modifications of a progressive wave caused by: (A) frictional damping, (B) convergence and corresponding effects produced by deepening, (C) and narrowing, (D) in a hypothetical estuary. (Reproduced and modified by permission of Total-Compagnie Francaise de Petroles.[1])

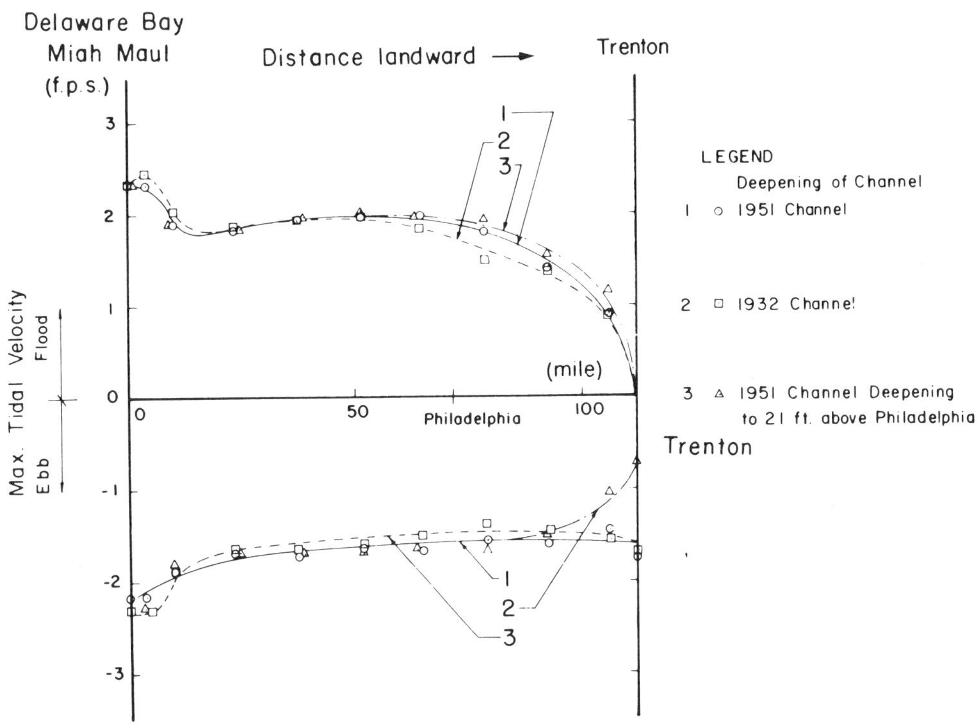

FIGURE 2. Effect of channel deepening on maximum tidal velocities along the length of the Delaware Estuary. Conditions 1 and 2 represent the distribution of velocity at various base conditions whereas condition 3 is with a deepened channel.[5]

simulations revealed early arrivals of high and low water, and an increase in maximum tidal velocity (Figure 2). Dredging in the Ems Estuary, Netherlands, between 1960 and 1980 reportedly[5] changed the tidal propagation by diminishing effects of bottom friction. After the channels were deepened and extended landward, the time lag between high water and high slack current at the head became shorter by about 9 min. Additionally, the tidal range at the head increased by 30 cm and consequently, the current velocities increased and created new current patterns. In response to these changes, shoals shifted and the morphology changed locally. According to de Jonge,[6] the resulting erosion between 1960 and 1980 caused an increase in turbidity throughout the estuary.

Frictional dissipation can also affect the tide wave symmetry. Where water depth in an estuary does not appreciably exceed the tidal amplitude, the tide wave is propagated as a shallow water wave whereby the speed, C, is given by:

$$C = \sqrt{gh} + \eta \tag{2}$$

where h is the water depth; g is the acceleration of gravity; and η is the local wave height of the tidal wave surface above still-water level. As the water depth decreases, η/h increases; hence, the tide wave speed decreases. Consequently, the flood crest (high water) propagates faster than, and tends to overtake, the trough (low water). This effect deforms the tide wave whereby the duration of rise (usually flood) is shortened and the fall (usually ebb) is prolonged. The resulting increase in flood velocities over ebb velocities can cause a strong landward dominance of sediment transport.[7] When dredging increases water depth and other factors are held constant, the rate at which the tide wave travels upstream increases. The wave profile, therefore, can loose the asymmetry that favors landward dominance of flood velocities in the deepened reaches.[3]

In contrast to frictional damping that decreases tidal amplitude landward, a decrease in channel cross section with distance upstream can cause a concentration of energy and thus an increase in amplitude (Figure 1B).[1] The relationship is given by Green's law, which ignores friction:

$$h = k\,(H^{1/4}\,B^{1/2} \tag{3}$$

where h is the wave amplitude; H and B are the channel depth and width, respectively; and k is a dimensional constant. Consequently, as an estuary channel narrows and shoals landward, which is the usual case, the tidal amplitude will increase. This trend opposes the effect of friction that reduces amplitude.

B. Equilibrium Concept

An ideal estuary channel is funnel shaped. As the banks converge and the bed shoals landward, widths, depths, and cross-sectional area diminish landward. When a progressive tidal wave is propagated landward from the ocean, its amplitude tends to be preserved. Therefore, for frictional dissipation to balance amplitude increase, there is an appropriate degree of convergence. That is, the rates of convergence of depth and width tend to balance frictional dissipation and thus conform to an exponential rate of change with minimal deviation.[8] To attain dynamic equilibrium, an estuary co-adjusts its tidal discharge and its channel geometry through erosion and deposition, or by changes in tidal characteristics including tidal wave length, amplitude, and the longitudinal gradient of tidal discharge. When an estuary is dredged to depths greater than those dictated by the equilibrium regime, sediments accumulate to reestablish an equilibrium depth in accord with the tidal hydraulics.

C. Models of Tidal Propagation

The variations of friction, convergence, and associated tidal amplitude are part of a spectrum of channel form and process combinations that display three models (Figure 3).[1]

1. *Convergence exceeds frictional dissipation.* If the loss of energy by friction is less than the convergence, the tidal amplitude increases landward into the estuary before decreasing toward the river, a trend characterizing a *hypersynchronous* estuary; *e.g.*, the Gironde Estuary, France.

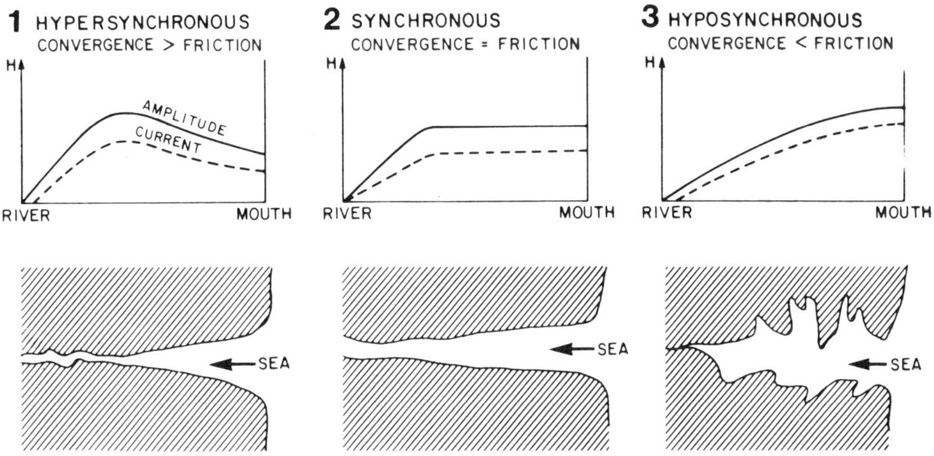

FIGURE 3. Modification of tidal range and current in estuaries with varying ratios of convergence to friction effects. (Reproduced and modified by permission of Total-Compagnie Francaise de Petroles.[1])

2. *Convergence equals friction.* The tidal amplitude is maintained before diminishing toward the river, a trend characterizing a *synchronous* estuary.
3. *Convergence is less than friction.* The tidal amplitude decreases landward into the estuary, a trend characterizing a *hyposynchronous* estuary; *e.g.*, the Hooghly River estuary, India.

These models apply only in cases where the tidal discharge is substantially greater than river discharge. Most estuaries are hypersynchronous, and tidal currents attain maximum strength in middle or landward reaches. Intensification of tidal currents causes bed erosion and resuspension of bed sediment. In the Gironde estuary, these processes produce a tidal turbidity maximum in upper reaches.[9] In the upper Delaware River estuary where dredging has removed shoals and rock, the effects of friction are reduced so that convergence exceeds friction. Consequently, tidal amplitude increases toward the estuary head.[10]

D. Standing Wave Effects

In some estuaries standing-wave properties are more important than progressive-wave properties. In a standing wave, the wave form remains stationary while the water flows rapidly through it (Figure 4A).[1] A standing-wave can form when an advancing progressive wave is reflected from an estuary shoal, bank, or head. Reflection occurs when the estuary length or distance between the mouth and shoal is a multiple of one quarter the wave length. Whereas reflection increases tidal amplitude over the shoal, removal of a shoal or bank by dredging can shift the position of the nodal point (Figure 4C). Consequently, tidal amplitude can be reduced seaward of the shoal but increased landward of the bank (Figure 4C). Wave energy that is not reflected can pass farther landward of the shoal and transform into an asymmetrical progressive wave (Figures 1B and 1D).

E. Tidal Discharge and Channel Stability

The bottom geometry of estuarine tidal channels continually changes in response to the transporting ability of the tidal discharge and the supply of sediment from external sources. Like fluvial channels, a tidal channel develops toward a state of maximum stability.[8] The channel must be neither too deep or too shallow for the amount of discharge and for the load of sediment that passes through it. To attain stability, an estuary co-adjusts its cross-sectional area to the discharge by erosion or deposition. The discharge, Q, is dependent on mean velocity, V, and cross-sectional area, A, so that at any section x:

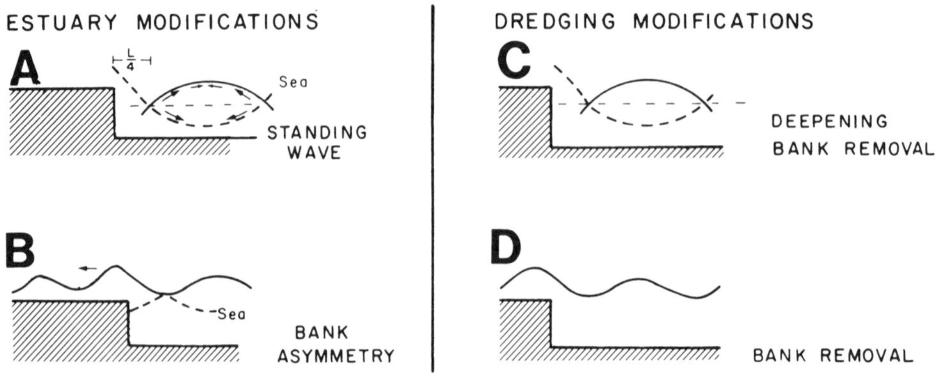

FIGURE 4. Modifications of a standing wave caused by: (A) reflection and (B) bank asymmetry and corresponding effects produced by dredging (C), (D) in a hypothetical estuary. (Reproduced and modified by permission of Total-Compagnie Francaise de Petroles.[1]) Figure 4C from OCEAN DUMPING AND MARINE POLLUTION; by H. Palmer and M. G. Gross; copyright (c) 1979 by Hutchinson and Ross. All rights reserved.

$$Qx = Vx\ Ax \tag{4}$$

Additionally, continuity must be maintained. Hence, longitudinal gradients of tidal discharge, rate of water-level change and channel width, must co-adjust to maintain a constant balance. In an equilibrium estuary, the total sediment transport rate over the channel cross section is proportional to the cross-sectional mean flow velocity. The dependence of estuarine entrance and of tidal inlet cross sections on maximum tidal discharge is confirmed by numerous studies of sandy channels.[11-12]

When an estuary is in long-term equilibrium, erosion balances sedimentation, but when dredging increases the water depth or width greater than those dictated by equilibrium, sedimentation sets in.[13] A small change in geometry of an estuary can produce large effects.

F. Case Histories

The effects of dredging and diking vary widely in different types of estuaries. In the Delaware Estuary, cross sections have been deepened by dredging and narrowed by landfill and diking, section by section, since 1850.[4] As shown in Figure 5, the cross-sectional areas along the estuary length in 1878 increase exponentially seaward suggesting quasi-equilibrium between estuary geometry and tidal discharge.[14] However, as dredging increased the depth and channel size, sedimentation rates increased in the enlarged sections. As training walls and dikes were constructed along margins of the estuary, partly to retain dredged material, tidal currents were trained into the channel axis of narrowed sections.[4] In these sections tidal currents eroded the channel to greater depths in sections without dikes, and thus sediment deposition was diverted from one section to another. This sequence of changes is indicated by deviations of cross-sectional areas from an exponential trend between 1878 and 1970 (Figure 5). Additionally, it is reflected in the formation of shoaling zones in some of the enlarged reaches.

In the Lune Estuary, U.K., a meandering, macrotidal estuary with a tidal range of about 7 m, the ebb current formerly moved from bank to bank through a low water channel whereas the flood current was relatively straight.[3,7] The natural equilibrium was upset in 1847 by dredging and by construction of training walls to suppress lateral movement of the channel.[7] The effects were large and immediate. At first the trained section became deeper; the tide wave moved into the estuary faster and the flood tide duration increased by 25 min.[3] The low water level became lower and the tide range increased. This trend produced greater

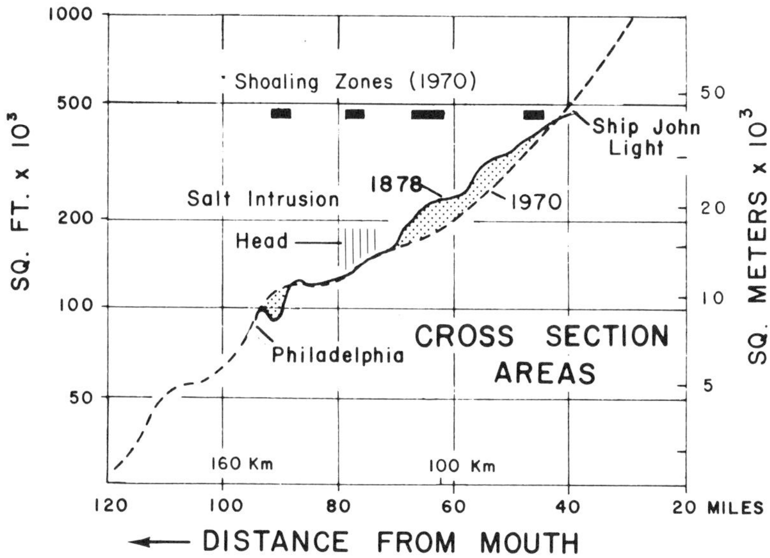

FIGURE 5. Changes in cross-sectional area along the Delaware Estuary between 1878 and 1970 in relation to shoaling zones in 1970. Zone of reduced cross section, shaded. (Reproduced with permission of van Nostrand Reinhold Co.)

flow velocities on the ebb tide and led to greater channel erosion. However, increased flood current behind the walls brought in a greater amount of sediment eroded from seaward areas. Consequently, massive sedimentation developed extending seaward from the estuary head. This led to a decrease in tidal volume and a decrease in ebb flow. The peak ebb and flood discharges were reduced by 47 and 20%, respectively. This promoted further sedimentation in the ebb channel. By 1955 the channel became relatively stable in the middle and upper reaches reflecting greater proportions of cohesive sediment and suppression of lateral erosion by the training walls.[3]

Dredging and jettying in the Seine Estuary, France, produced large geometric changes with marked consequences.[15] Prior to 1850, the estuary consisted of shallow and meandering ebb and flood channels bordered by extensive tidal flats (Figure 6A). Sedimentation rates were relatively low. To accommodate larger ships during the 19th century, the channel was dredged from a natural depth of about 3 to 8 m, and it was stabilized and straightened by diking and jettying (Figures 6B and 6C). Additionally, intertidal flats and marshes were reclaimed for industrial sites, thus reducing the volume of the estuary almost 50%.[15] These alterations concentrated river discharge and ebb flow into a narrow, jettied channel that amplified current velocities locally. In contrast, currents behind the jetties became weaker, and this induced rapid sedimentation on the tidal flats. As the tidal flats and marshes grew upward to mid and high tide levels, the area available for sedimentation to the middle estuary was reduced. The locus of sedimentation in the channel, therefore, shifted seaward and necessitated extending the jetties farther seaward. In turn, this produced greater current amplification and seaward advection in the main channel. Both the landward limit of salty water and the turbidity maximum shifted seaward (Figure 7). As a result, the turbidity maximum now extends out of the estuary during periods of high river-flow and spring tide. This allows suspended sediment to escape from the estuary while deposition increases the size of offshore mud zones. Over the long-term, dredging and jettying have thus changed the role of the estuary from a sink for fluvial and marine sediment to a source of fluvial sediment for the shelf.[15]

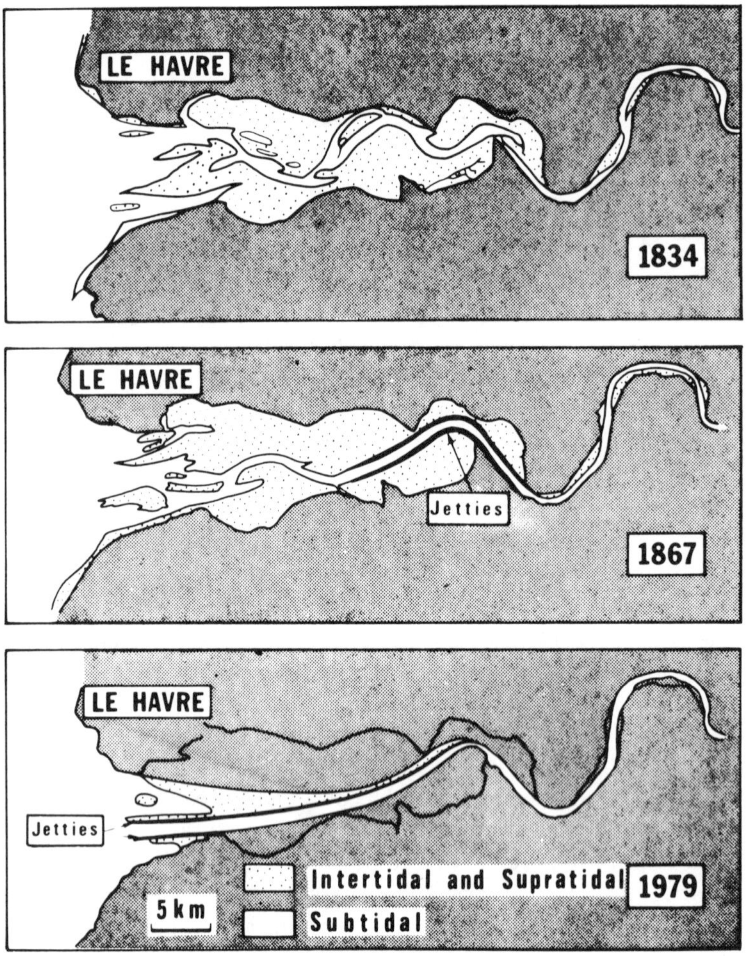

FIGURE 6. Change in morphology of the Seine Estuary between 1834, 1867, and 1979 showing change in channel configuration and distribution of intertidal flats and marshes. (Reproduced with permission of Elsevier Scientific Publishing Company.)

III. EFFECTS ON THE ESTUARINE CIRCULATION

Channel depth is one of the important parameters that control the sequence of estuarine mixing types.[16] If river flow, tidal velocity, and width are held constant, then an increase in depth lowers the effectiveness of tidal velocities in promoting vertical mixing between lower and upper estuarine layers. Consequently, haline stratification increases and the net volume transport in each layer is reduced. Thus, river outflow is more confined to the upper layer whereas salty inflow passes upstream more effectively through the lower layer. The overall effect of deepening is to shift the mixing type from a Pritchard Type C or B toward Type A.

As dredging removes entrance shoals or bars, and as deepening increases the volume of the channel, the ability of river flow to hold back saltwater is reduced. In zones of high salinity gradient, both the shear stresses and the horizontal pressure gradients are reduced.[3] Consequently, saltwater penetrates farther landward than its normal position, and this leads to a corresponding landward movement of the null zone (Figure 8A). Since transported sediments tend to accumulate in the null zone forming a turbidity maximum or shoals, the locus of these features can also shift landward.

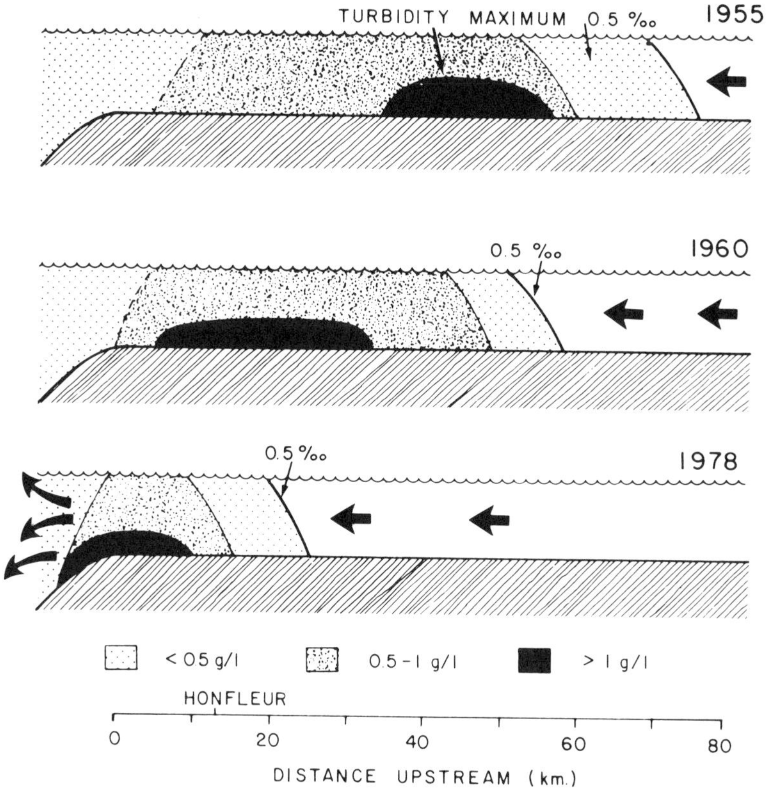

FIGURE 7. Seaward migration of the turbidity maximum and the salt intrusion at 0.5‰ in the Seine Estuary between 1955 and 1978. (Reproduced with permission of Elsevier Scientific Publishing Company.)

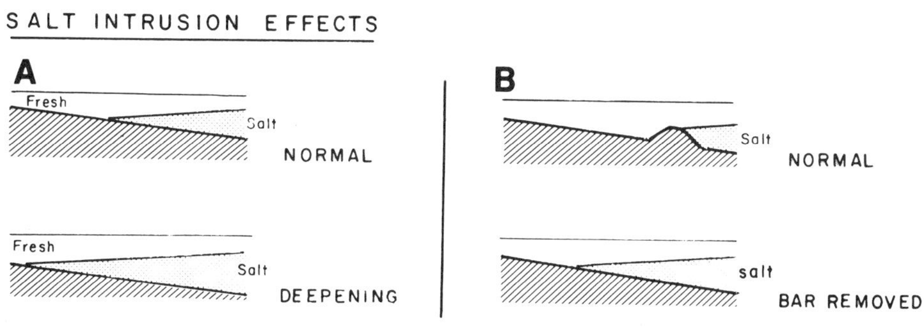

FIGURE 8. Schematic diagram showing (A) effects of channel deepening and (B) effects of removing an entrance bar, on penetration of the salt intrusion.

In a two-dimensional, steady-state numerical model, Festa and Hansen[17] found that salinity stratification decreased as depth increased. This trend is true only near the mouth of an estuary; elsewhere an increase in depth increases the stratification.[18]

A. Case Histories

In Savannah Harbor, a partly mixed estuary, the channel has been progressively deepened from about 3.7 m in 1891 to 11.4 m in 1975, and also extended landward. Maintenance

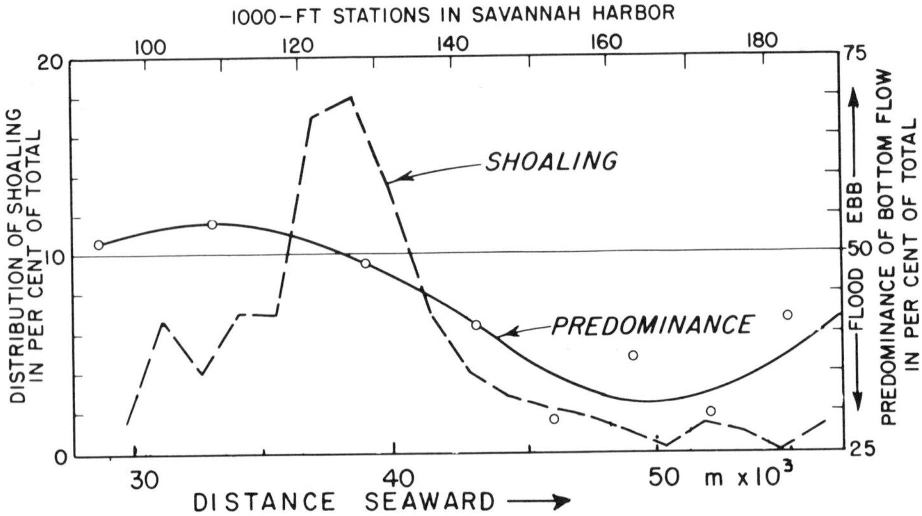

FIGURE 9. Relation of sediment shoaling to predominance of bottom flow in the Savannah Harbor estuary.[19]

dredging records for the 11 m channel show that of the total dredged material, about 5.3 $\times$ 10^6 m^3/year, more than two thirds was dredged from a 9 km long zone.[19] This shoaling zone lies in the vicinity of the null zone for bottom flow predominance (Figure 9), i.e., a zone where the bottom flow changes from landward to seaward. The effects of successive channel deepening between 1923 and 1953 were twofold. First, the sedimentation rates in the estuary increased drastically, from 2.1 to 5.5 $\times$ 10^6 m^3/year. Second, the locus of maximum sedimentation shifted landward 19 km, a zone where disposal of dredged material was difficult. With additional dredging between 1953 and 1975, the increase in harbor-wide shoaling was relatively small. Reportedly,[20] by the time of the final channel deepening, the density circulation had developed to such a degree that most potential sediment entering the harbor was trapped.

Dredging in South Pass and Southwest Pass of the Mississippi River has increased the depth of entrance bars from about 3 to 12 m. Prior to deepening, saltwater was largely limited to the vicinity of the bars (Figure 8B). After deepening, however, saltwater penetrated upstream over 200 km at extreme low river inflow.[20] New bars formed in the landward tip of the salt wedge far upstream of the entrance. If the channel depth were allowed to return to its natural state, reportedly,[20] the new bars would build up with time, density currents would be reduced and the bars would gradually shift seaward and reform in their original position.

The effects of a proposed 3 m channel deepening in sections of the James River estuary were studied in a hydraulic model to predict changes that might affect oyster production.[21] Deepening produced the greatest salinity change where the major cut was performed. The lower layer in the channel with salinities of about 10 to 16‰ became saltier by about 0.5‰ whereas the upper layer became fresher by about 0.3‰. Salinity changes were greatest at intermediate river inflow. As saline stratification increased, tidal velocities were less effective in vertical mixing between estuarine layers, and the net volume transport in each layer was slightly reduced.

IV. CONCLUDING COMMENTARY

The foregoing concepts and case histories emphasize how dredging can provoke significant changes in the dynamic behavior of an estuary and thus, create an imbalance between erosion and sedimentation. The further a system deviates from its natural equilibrium state, the greater the imbalance and the problem becomes worse.[22,23] As channels are dredged deeper, the potential for entrapment of sediment is increased. Greater sedimentation necessitates more dredging and therefore dredging is self-perpetuating.

REFERENCES

1. **Salomon, J. C. and Allen, G. P.,** Role sedimentologique de la maree dans les estuaries a fort marnage, *Compagnie Francais des Petroles, Notes and Memoires,* 18, 35, 1983.
2. **Ippen, A. T. and Harleman, D. R. F.,** Tidal dynamics in estuaries, in *Estuary and Coastline Hydrodynamics,* Ippen, A. T., Ed., McGraw-Hill, New York, 493, 1966.
3. **McDowell, D. M. and O'Connor, B. A.,** *Hydraulic Behavior of Estuaries,* John Wiley & Sons, New York, 1977, 292.
4. **Pillsbury, G. B.,** *Tidal Hydraulics,* Prof. Paper of the Corps of Engineers, No. 34, U.S. Government Printing Office, Washington, D. C., 1955, 247.
5. **Harleman, D. R. F. and Lee, C. H.,** *The Computation of Tides and Currents in Estuaries and Canals,* Tech. Bull. 16, Comm. on Tidal Hydraulics, U.S. Army, Corps of Engineers, 1969, 276.
6. **De Jonge, V. N.,** Relations between annual dredging activities, suspended matter concentrations and the development of the tidal regime in the Ems Estuary, *Can. J. Fish. Aquatic Sci.,* 40 (Suppl 1), 289, 1983.
7. **Inglis, C. C. and Kestner, F. J.,** The long-term effects of training walls, reclamation and dredging on estuaries, *Proc. Inst. Civ. Eng.,* 9, 193, 1958.
8. **Wright, L. D., Coleman, J. M., and Thom, B. G.,** Processes of channel development in a high-tide-range environment: Cambridge Gulf-Ord River Delta, *J. Geol.,* 81, 15, 1973.
9. **Allen, G. P., Salomon, J. C., Bassoulet, P., Du Penhoat, Y., and De Grandpre, C.,** Effects of tides on mixing and suspended sediment transport in macrotidal estuaries, *Sediment. Geol.,* 26, 69, 1980.
10. **Wicker, C. F.,** *Evaluation of Present State of Knowledge of Factors Affecting Tidal Hydraulics and Related Phenomena, Rept. No. 3,* Comm. on Tidal Hydraulics, U.S. Army Corps of Engineers, 1965, 223.
11. **O'Brien, M. P.,** *Notes on Tidal Inlets on Sandy Shores, Rept. 5,* U.S. Army, Coastal Engr. Res. Center, GITI, 1976, 20.
12. **Bruun, P. and Gerritsen, F.,** *Stability of Coastal Inlets,* North-Holland, Amsterdam, 1960, 123.
13. **Inglis, C. C. and Allen, M. A.,** The regimen of the Thames Estuary as affected by currents, salinities, and river flow, *Proc. Inst. Civ. Eng.,* 7, 827, 1957.
14. U.S. Army Engineer District, Philadelphia, Corps of Engineers, *Long Range Spoil Disposal Study,* Part III, 1973, 140.
15. **Avoine, J., Allen, G. P., Nichols, M., Salomon, J. C., and Larsonneur, C.,** Suspended-sediment transport in the Seine Estuary, France: effect of man-made modifications on estuary-shelf sedimentology, *Mar. Geol.,* 40, 119, 1981.
16. **Pritchard, D. W.,** Estuarine circulation patterns, *Proc. Am. Civ. Eng.,* 81, 717, 1955.
17. **Festa, J. F. and Hansen, D. V.,** A two-dimensional numerical model of estuarine circulation: the effects of altering river discharge, *Estuarine Coastal Mar. Sci.,* 4, 309, 1976.
18. **Pritchard, D. W.,** Estuarine classification — a help or hindrance, in *Estuarine Circulation,* Neilson, B., Kuo, A., Brubaker, J., Eds., Humana Press, Clifton, N.J., 1987.
19. **Simmons, H. B.,** *Channel Depth as a Factor in Estuarine Sedimentation,* Tech. Bull. 8, U.S. Army Committee on Tidal Hydraulics, 1965, 15.
20. **Simmons, H. B. and Herrmann, F. A.,** Effects of man-made works on the hydraulic, salinity, and shoaling regimens of estuaries, in *Environmental Framework of Coastal Plain Estuaries,* Nelson, B. W., Ed., Geological Society of America Memoir, Boulder, Co., 133, 555, 1972.
21. **Nichols, M.,** Effect of increasing depth on salinity in the James River Estuary, in *Environmental Framework of Coastal Plain Estuaries,* Nelson, B. W., Ed., Geological Society of America Memoir, Boulder, Co., 133, 571, 1972.
22. **Nichols, M.,** The problem of misplaced sediment, in *Ocean Dumping and Marine Pollution,* Palmer and Gross, Eds., Dowden, Hutchinson, and Ross, Stroudsburg, Pa., 1978, 147.
23. **Price, W. A. and Kendrick, M. P.,** Dredging and siltation — cause and effect, in *Proc. I.C.E. Symposium on Dredging,* Inst. of Civil Engr., London, 1967, 31.

Chapter 18

DJINNANG II: A FACILITY TO STUDY MIXING IN STRATIFIED WATERS

Jörg Imberger and Richard Chapman

TABLE OF CONTENTS

I. INTRODUCTION

Most inland lakes, estuaries, and coastal waters are characterized by a well-defined density stratification of the water column.[1] This stratification may be introduced by a riverine buoyancy flux or may be due to solar heating. In most Mediterranean climates, the stratification persists for a large part of the year, stabilizing the water column and suppressing mean vertical motions and active turbulence. Surface stresses introduced by the winds are thus insufficient to completely mix the water column; instead, a surface mixed layer forms where most of the introduced kinetic energy is dissipated. Below this surface layer only weak, sporadic mixing events may be observed.

The study of mixing in lakes and coastal regimes must recognize the patchiness, introduced by this stratification, of any turbulence present in the water column. For instance, in wind-induced, surface-mixed layers[2] or gravitational overflows[3] the turbulence field is partitioned vertically by a sharp density interface, and horizontally by frontal structures.[4] Similarly, in the hypolimnion, topographic gyres,[5] basin scale internal waves[6] or shelf waves,[5] all sustain smaller scale mixing patches below the surface mixed layer.[7] Mixing within an active region may itself be divided into entrainment motions and small scale mixing motions;[8] entrainment is the action of coherent structures engulfing neighboring fluid, while mixing is caused by small-scale instabilities of the interleaving engulfed fluid layers. The subsequent collapse of an active entrainment event leads to further internal wave motions and the formation of additional intrusions;[7] these intrusions move horizontally very slowly, redistributing the potential energy gained locally during the mixing event. Rotation is normally thought to influence only basin scale motions, but it is not difficult to show[9] that the internal radius of deformation of these intrusions may only be tens of meters; an intrusion may thus have a thickness of 1 m at one boundary and be almost of zero thickness at the other boundary of the basin.

The assessment of the rate of mixing depends on two major questions. First, given the spectrum of mean circulation currents, slow intrusions, and internal waves, how do these statistically combine to form the fine scale velocity field responsible for inducing the observed entrainment motions? Second, given a particular fine scale velocity field, observed to induce an entrainment motion, how much mixing (change of potential energy) results from such an isolated event?

In this context, we shall define four scales of motion. First, the basin scale motion and oscillations; second, the fine scale motion setting the stage for an entrainment event; third, entrainment motion (Kelvin-Helmholtz billows, wave breaking, boundary layer turbulence, critical layer absorption); and fourth, scales associated with the mixing of mass and mechanical and thermal energy.

Similarly, there is a hierarchy of timescales. Characteristic times for basin or shelf scale motions are measured in hours. Fine scale velocities have timescales as short as tens of seconds and entrainment motions have a life not much longer than a few seconds, while the mixing timescales are measured in tens of milliseconds.

The task is thus formidable! In order to establish the casual relationships for the physical dynamics requires the measurement of velocity, temperature, and conductivity on a space scale from kilometers (to measure the energy input) to fractions of a millimeter (to measure the energy dissipation). Temperature and salinity gradient dissipation measurements require resolution down to hundredths of a millimeter. Compounding this spatial scale range requirement is a requirement for instrument time response down to tens of milliseconds. Perhaps the greatest difficulty in measurement is, however, brought about by the nonhomogeneity, intermittency and nonstationarity of the turbulent field at a point. Under such circumstances, single point moorings contribute little toward answering the global mass and momentum transfer questions.

However, it must again be stressed that in a strongly stratified fluid the active turbulence is nearly always confined to narrow fronts, gradient regions or isolated patches. The bulk of the fluid is either in a quiescent nonmixing state or in a near homogeneous turbulent state as in a surface mixed layer. Mixing events tend to be isolated and mostly noninteracting, distinct from the coherent motions in active turbulence. There is thus some hope of isolating and identifying a particular entrainment motion and to measure or model the subsequent mixing.

The research facility Djinnang II (Aboriginal "to discover") was specifically designed to allow the undertaking of such a measurement program. It is an evolving design and future changes are contemplated. However, having reached stage II, reasonable design criteria have been identified and these are described in this paper.

The system may be conveniently thought of as consisting of four components. First, the basin scale motions and meteorological forcing are measured using *in situ* weather stations, thermistor, and conductivity chains, as well as a portable CTD mounted on a small dinghy. This information is augmented by measurements from a remote profiling station capable of measuring velocity, temperature, conductivity, visible light, and some turbulent properties of the water column. Second, intense CTD work, directly from the Djinnang II, is used to define the fine scale spatial variability. Recent efforts (see also Veronis[10]) have shown that the inverse technique may also be used to estimate the stream function and mean diffusion coefficients from this type of data. If successful this will allow insight to be gained into the relationship between the mean mass and momentum fluxes and the instantaneous values measured directly from the profiles. Third, entrainment and mixing events are synoptically identified in the field with remote acoustic sensing. Fourth, information from both high resolution CTD data and data from temperature and salinity microstructure vehicles are being used to estimate the energy dissipation, vertical diffusion coefficients, and indicators of the degree of anisotropy in the turbulent field.

In order to maximize the likelihood of success in isolating and measuring properties of the entrainment motions, the facility is designed to allow on-line processing. All data from the *in situ* stations are sent by telemetry to the central system on the Djinnang II where they are combined with data collected from side casts, perimeter CTD data, and data from the microstructure vehicles. A series of diagnostics are calculated and displayed in graphical or contour form to assist the interpretation and aid the decision making scientist onboard and to guide the deployment of the facility. Results from the numerical simulations may also be overlayed with the actual measurements.

II. *IN SITU* INSTRUMENTATION

A. Meteorological Stations

The meteorological sensors (Table 1) are normally mounted on a guyed tower erected in the water column. The mast allows deployment in up to 30 m of water (Figure 1a). The data are recorded digitally by a remote data logger designed around two CMOS national NSC800 microprocessors. Static RAM is used for storage and the system is powered by solar charged sealed lead acid batteries.[11] The data logger has provision for 16 analog channels, 3 interrupt channels, 2 frequency channels, and 16 digital channels. The use of two processors allows communication between the data logger and the telemetry unit while data acquisition is proceeding. Further, extensive computations can be carried out in the remote data loggers so that the required fluxes and other correlations may be calculated from rapidly sampled raw data, thus reducing storage requirements. Floating point arithmetic is used throughout to avoid problems with dynamic range. The telemetry is organized with a poled retransmit on error protocol so that a large number of remote stations can be serviced by a single VHF channel and a single telemetry master station situated on the Djinnang II.

Table 1
SENSOR SPECIFICATIONS

Manufacturer	Model	Beamwidth	Frequency	Power	Sensitivity
			Acoustic Transceivers		
Datasonics	DFT210	22°	32 KHz	4KW	.5μV @ 50Ω
Ross	4200	3.5°	100 KHz	1KW	.5μV @ 50Ω
Datasonics	DFT210	2.0°	300 KHz	4KW	.5μV @ 50Ω

Sensor	Manufacturer	Model	Resolution	Accuracy	Time Constant
			Finescale Sensors		
Conductivity	Seabird (pumped)	SBE-4	—[a]	—[a]	2×10^{-2} sec
Temperature	Seabird	SBE-3	10^{-3}°C	10^{-2}°C	7×10^{-2} sec
Depth	Paroscientific	8060 DS	10^{-3} m	3×10^{-3}m	—
pH	Orion	81-02	10^{-3} pH	—	2.5×10^{-1} sec
Velocity	Neil Brown	ACM-2	—	10^{-2} m sec^{-1}	2×10^{-1} sec
Tilt	Schaevitz	LSRP	10^{-3} radian	10^{-3} radians	8×10^{-3} sec
Orientation	Neil Brown	ACM-2	.04	2°	2×10^{-1} sec
Solar Radiation	LI-COR	LI-190SEB	—[a]	—[a]	—
Lift Sensor	Under Sea Technology, Inc.	Shear Probe 83-1001	—	—	—
			Microscale Sensors		
Temperature	Thermometrics	FP07	6×10^{-4}°C	10^{-2}°C	1.1×10^{-2} sec
Temp Grad.	—	—	10^{-3}°C m^{-1}	2%	1.1×10^{-2} sec
Conductivity	P.M.E.	106	4×10^{-4} Sm^{-1}	1.2×10^{-1} Sm^{-1}	4×10^{-3} sec
Cond Grad.	—	—	10^{-2} Sm^{-2}	2%	4×10^{-3} sec
Depth	Keller	PAA-10/10	2×10^{-3} m	10^{-1}m	—

		Boat Instruments		
Position	Motorola	Mini Ranger III	1 m	
Heading	Western Marine	DC 710-1000	1°	
Depth	Western Marine	DOC 310	10^{-1} m	
Speed	Sumlog	Sumlog II	5×10^{-2} m sec^{-1}	
		Meterological Sensor		
Temperature	Veco	41A32	10^{-2}°C	10^{-2}°C
Wind	Weathermeasure	173A	6×10^{-2} m sec^{-1}	± 3%
Humidity	Vaisala	HMP14U	—	± 3%
Net Radiation	Schenk	8110	.5 W m^{-2}	± 3%
Radiation	Matrix Inc	MK1-G	.5 W m^{-2}	± 5%
Barometer	Setra Systems	270	2×10^{-1} mb	± .1%
Rain	Rimco	R/TBR	2×10^{-4} m	± 1%
		Mini Logger		
Temperature	Veco	41A32	10^{-2}°C	±0.1°C
Wind	Rimco	480640	0.1 m sec^{-1}	±0.1 m sec^{-1}
Rain	Rimco	R/TBR	2×10^{-4} m	±1%

[a] Resolution and accuracy of these sensors is a strong function of the operating point.

A

B

FIGURE 1. General photographs of a selection of the equipment. (A) weather station mast is shown at the left of photograph and the RPS with the associated logger are depicted at the right of the photo; (B) general view of the Djinnang II with a thermistor chain logger in forefront; (C) general view of the microstructure vehicle with drag plate; and (D) the F probe ready for deployment in the Djinnang II.

FIGURE 1C

FIGURE 1D

All communication packages utilize CRC16 block checking to ensure error-free operation of the telemetry system.

B. Mini Loggers

The mini loggers are small, economical data loggers easy to deploy, designed and developed to record frequency signals from a wind anemometer, a simple temperature thermistor circuit, or an acoustic depth measuring device (Table 1). In use they are a single channel device designed for single-point readings around the periphery of any study area. The logger construction is again based on a Z80 National NSC800 microprocessor chip supplemented with 3000 bytes of static RAM. Communication with the Djinnang II is via a "plug-in" serial line.[12]

C. Remote Profiling Station (RPS)

The remote profiling station was developed[13] after testing showed that a rapidly free-falling vehicle could not be stabilized sufficiently to allow measurement of the very low velocities normally found in lakes and estuaries. The station (Figure 1a) consists of a rigid guyed tower, of modular construction, which can be deployed in up to 30 m of water. The tower rigidly supports two accurately aligned rails onto which a trolley may be mounted to carry the desired instrumentation package. The trolley is under positive cable drive and the system is powered at the top of the mast by an electric motor. Power for the station is supplied from solar cells with sufficient lead acid battery storage to permit 20 profiles in the absence of solar radiation. The power supply unit, remote logger, and telemetry unit are mounted on a raft tethered at the mast. On command from the Djinnang II, the system executes a single profile, storing the data in the remote data logger before transmitting it to the Djinnang II on request. When transmission is complete, a new profile can commence.

The profile depth and frequency is controlled through the Djinnang II system, but the profiling speed at 1 m sec^{-1} is fixed.

D. Portable Data Acquisition System (PDAS)

The portable data acquisition system is a Z80-based microprocessor unit with dual flexible disk drives as a storage medium. The computer system is described in Scolaro and Luketina.[14] The unit is housed, together with a graphics terminal, in a water-resistant compartment suitable for deployment in a small dinghy. This portable device is used for general purpose field computing and when coupled with one of the CTDs provides an accurate instrument with flexible computer control. Software exists to allow on-line plotting of the cast information, editing of the file, and storage on a floppy disk drive for later retrieval.

III. FINE SCALE PROFILING CTD EQUIPMENT

At present, three CTDs are in use. Each one is designed for a special purpose and constructed around the same computer system as the portable data acquisition system. All three probes are fitted with the Sea-Bird temperature and conductivity sensors and the Paroscientific depth sensors (Table 1). The frequency data are counted and converted to physical variables in the probe system. Analog sensor signals are digitized in the probe and together with the frequency information are transmitted to the deck unit (Djinnang II, RPS or PDAS) via a coaxial cable operating at 38,400 bits per second. The sampling rate is set at 50 Hz to allow signal enhancement to correct for sensor response. In general the sensors were chosen for their accuracy and stability and not for their spatial resolution since the fine scale motions to be resolved with these sensors are generally larger than 0.10 m.

A. Portable P Probe

The P probe, of rugged construction, is configured with a depth, temperature, and conductivity sensor and is used mainly with the PDAS from a small dinghy. Sampling is software controlled with a conditional channel being optional. However, this instrument is most often deployed as a free-fall vehicle ballasted to fall at 1 m sec^{-1} with acquisition at 50 Hz.

B. The Fine Scale Probe F

The F probe is equipped with depth, temperature, conductivity, and pH sensors (Figure 1d). A dissolved oxygen sensor with matching resolution is presently being developed. The F probe is mechanically the most flexible and when used from Djinnang II may be deployed in three distinct modes. First, in the free-fall mode, with depth or time as a conditional channel; second, towed at a predetermined depth with time as the conditional channel; and third, in a YO-YO pattern yielding a vertical section through the water body. The sensor package of the F probe is pivoted at a knee joint and the package possesses a fin to allow the sensors to be directed into the flow under all deployment configurations.

C. The Mixed Layer Probe L

The L probe was designed[15,16] as a mixed layer sensor package and is equipped with a depth, temperature, conductivity, light intensity, and a two-axes acoustic current meter. In addition, the probe contains a lift turbulence velocity sensor (Table 1). An accurate compass and two sensitive tilt sensors inside the housing monitor the spatial orientation of the vehicle and allow deployment either as a free-fall vehicle or from the RPS. The system schematic is shown in Figure 2. It has been found that the free-fall mode is suitable for estuarine high velocity situations, but for slow motions, as normally found in a lake hypolimnion, it is necessary to attach the probe to the trolley of the RPS. When attached to the RPS the acoustic current sensors can resolve velocity down in the millimeter per second range.

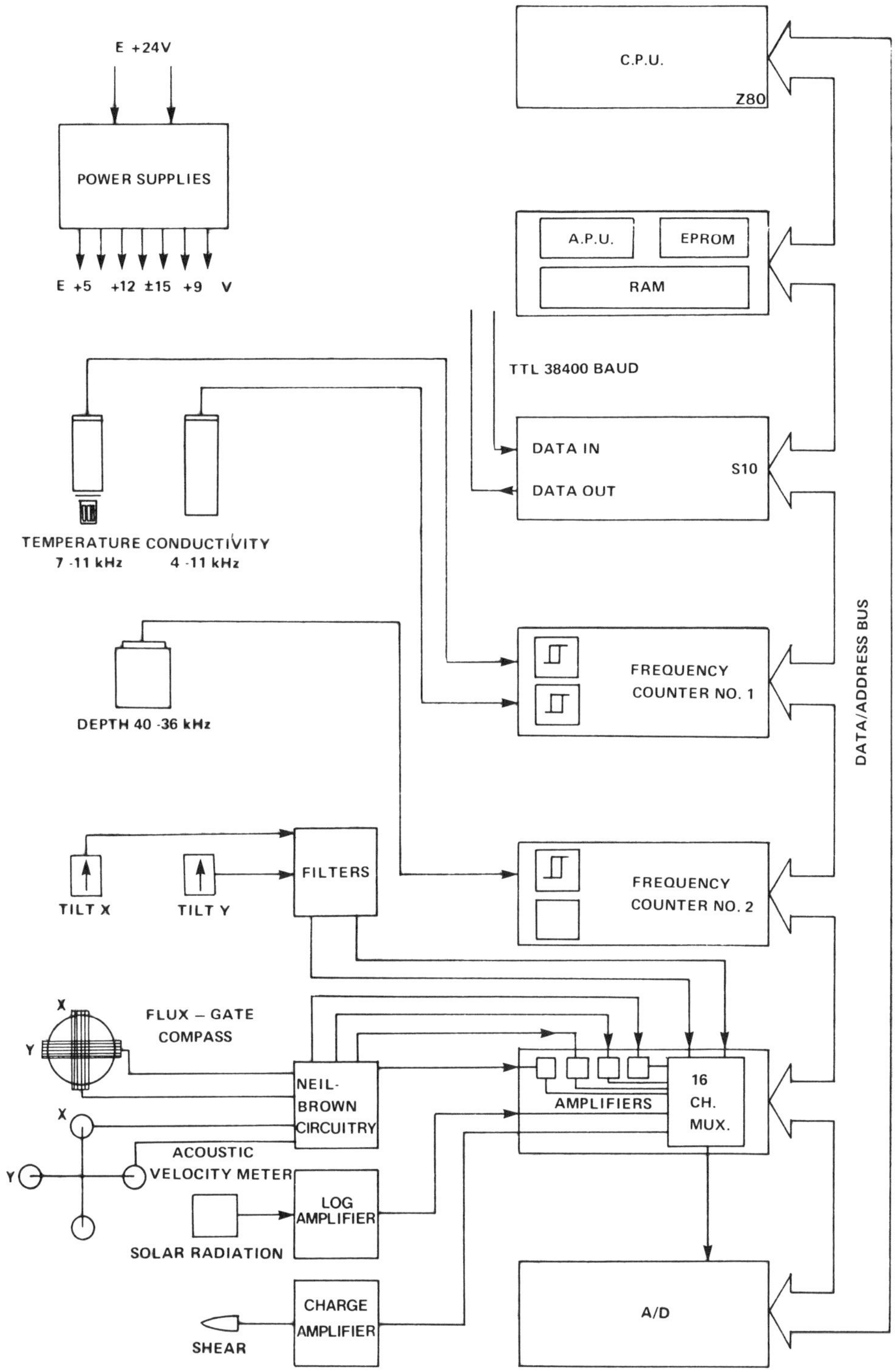

FIGURE 2. Schematic of the mixed layer probe electronics.

IV. MICROSTRUCTURE VEHICLES

The term ''microstructure'' refers to the changes in a tracer field at the entrainment mixing scales. At the mixing scales it is reasonable to investigate the properties of turbulence and to compare these to signatures as predicted by the theory of homogeneous isotropic turbulence. The logic being that at such small scales buoyancy or rotation do not influence the motion, and both the temperature and conductivity fields behave as pure tracers. Caldwell et al.[17] have shown that this assumption is a useful concept and much can be learned by measuring the spectral properties of the temperature and salinity gradient field down to the dissipation cut-off length $0(\nu\kappa^2/\epsilon)^{1/4}$. At present, using a fast tip thermistor, temperature gradients may be used for dissipations up to 10^{-6} m^2 sec^{-3},[18-20] however, in the case of salinity, the small diffusivity of salt reduces this dissipation maximum to 10^{-10} m^2 sec^{-3}, making it of limited use for dissipation measurements. The conductivity sensor is most useful as a fast response temperature indicator[20] in a uniform salinity ocean and also in calculating the salinity and density displacement lengths of the entrainment motions a good indication can be achieved of the rate of turbulent collapse in strongly stratified conditions.

An initial set of instruments were constructed, patterned on the original Oregon State University design,[21] with an umbilical cord and analog signal transmission. However, while most useful information was obtained with these instruments, they had two major drawbacks. First, the umbilical cord made rapid deployment very difficult and so only a single pass could be obtained through any particular overturn event, thus defeating the aim of the system as a whole. Second, the analog transmission was prone to interference from the acoustic imaging system, and the two could not be used together. It was therefore decided to develop a completely new self-contained free rising microstructure profiler (Figure 1c), able to measure temperature and conductivity microstructure. A version containing a set of 2 sensors is shown in Figure 3. The vehicle[22] has provision for 3 temperature and 3 conductivity channels and is serviced with a 16-bit A to D converter and a National Semiconductor NS16032 microprocessor. This configuration overcomes the dynamic range problems of 12-bit conversion, yields large address capability, and the processor is fast enough to allow on-line computation of spectral estimates. The latter property reduces the storage requirements (1 megabyte of dynamic RAM) making the vehicle suitable for extended profiling.

The signals from the sensors are passed through analog differentiators, and one of the temperature gradient channels has dual amplification in order to allow the extraction of amplifier noise.[21] The signals are digitized at 100 Hz. The processor is serviced with a nonvolatile memory, and the programs may be downloaded from the Djinnang II. Various sampling strategies are being developed, but a most common deployment is to preset the depth at which the ballast weight is to be released; the vehicle loaded with the ballast is then jettisoned into the water. On the downward path the drag plates swivel, causing the vehicle to glide down at a 45° angle. At the preset depth the ballast is released on command from the processor, and the vehicle commences its ascent with the sensors facing upward. The vehicle rises at about 0.1 m sec^{-1} piercing the free surface, thus yielding good data to within a few millimeters of the free surface. On reaching the surface the vehicle is retrieved and coupled to an umbilical cord through which the data is transferred to the Djinnang II computers.

V. DJINNANG II

The Djinnang II is an 8-m, twin-hulled craft chosen for its stability and high speed (Figure 1b). The hull, body, and air-conditioned cabin are constructed of Kevlar fiber for lightness and durability. The vehicle is equipped with two 175 horsepower outboard motors and the equipment is powered from a 7KVA Onan diesel generator unit.

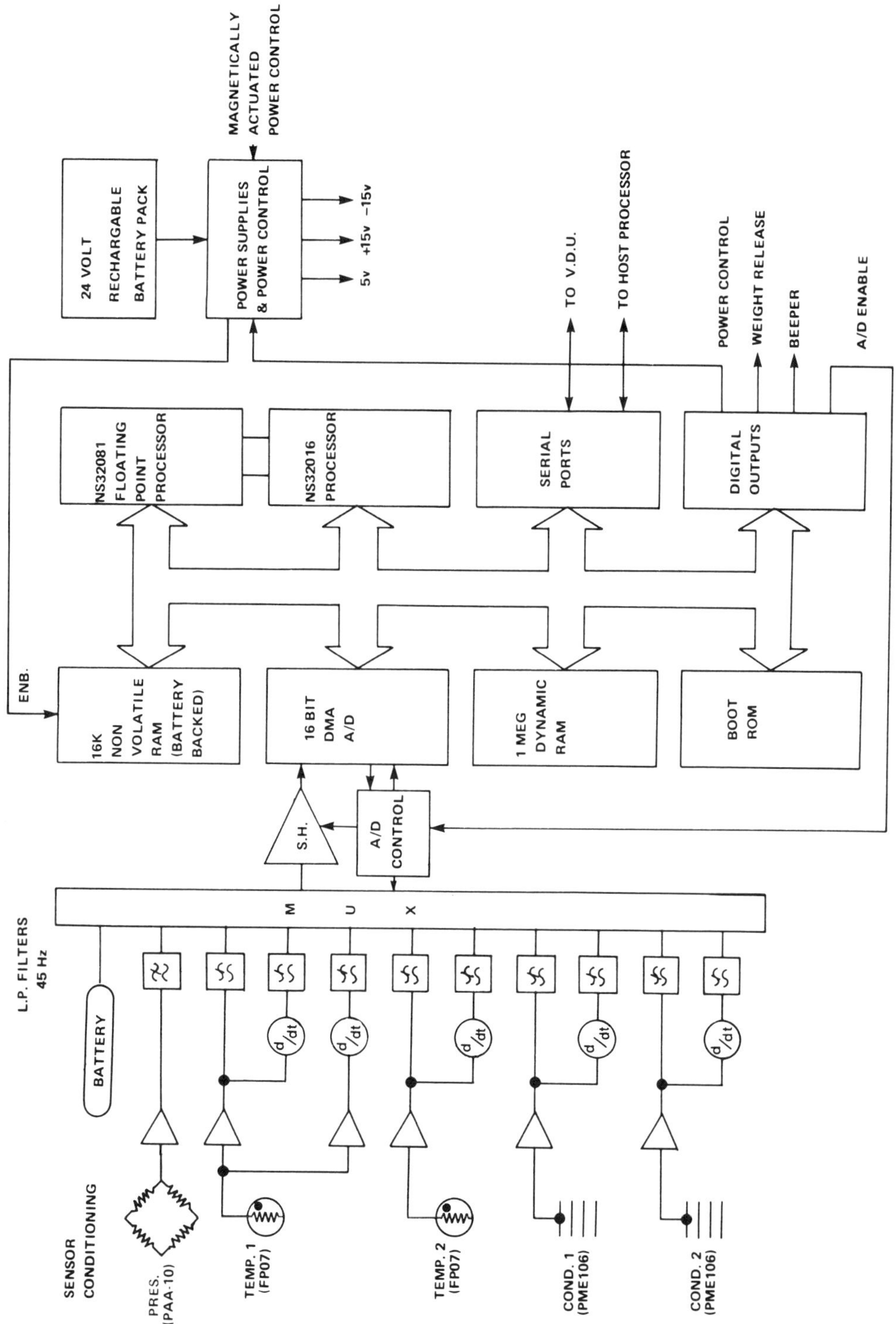

FIGURE 3. Schematic of the microstructure probe electronics.

The Djinnang II forms the central unit to the four sets of instrumentation; the *in situ* equipment, acoustic imaging system, fine scale profilers, and microstructure vehicles can all simultaneously feed their data into the Djinnang II acquisition and computing system (see Figure 4). These data are then combined with the boat position, boat speed, orientation of the boat, and depth of the water to form a single data set as described below.

A flexible system architecture has been devised to facilitate the ready interfacing of a wide variety of different instruments and computers. Dividing the system into well-defined subsystems enables independent operation of different parts of the system permitting upgrading of the various components with minimum disruption. Further, the subsystem concept adds a large degree of system independence and thus operational redundancy, yielding a total system with little down time.[23]

A. Data Concentrator

The data concentrator uses a Z80 microprocessor chosen for the simplicity of interfacing. In addition to supplying data to the HP1000, the concentrator also permits the Motorola 68000 based acoustic imaging system to request mini-ranger and date/time data.

B. Data Acquisition System

The data acquisition system uses an HP1000 computer. The system has sufficient speed to perform real-time calibration of all sensors, to provide real-time graphics display of the boat position, and to perform system operation diagnostics. The data concentrator and the data acquisition system communicate via a dedicated high speed parallel direct memory access bus.

The HP1000 system in the Djinnang II has proven to be exceptionally robust and reliable in several years of operation in the earlier vessel Djinnang. The use of only flexible disk storage allows it to operate in the most extreme environmental conditions. As it is dedicated to the task of data acquisition, it has no requirement for a large data base and is never committed to data processing. It is, therefore, possible to acquire data virtually continuously while still receiving feedback from the data processing system on the progress of an experiment. The principle functions of the HP system are as follows:

1. To provide a menu driven acquisition control facility, allowing the operator to select appropriate sampling regimes for a particular experiment
2. To prompt the operator for essential parameters such as location of mini-ranger reference stations
3. To provide a real-time display of all monitored variables before and during acquisition
4. To provide a real-time graphics display of boat position, to enable the helmsman to precisely locate predefined stations, or to accurately steer any required course
5. To synchronously sample up to 8 analog channels
6. To implement a sampling algorithm so that data are only stored at an appropriate density for a particular experiment
7. To perform calibration functions so that the real-time display can be easily interpreted
8. Store all data immediately on the flexible disks
9. Transfer the data to the data processing system at the completion of the data set, together with a directive as to what processing should be performed

The system has 320 K bytes of memory and runs the real-time memory based operating system RTEM. The user has indirect access from the HP terminal to all peripherals of the data processing system via a link. The link driver software is written so that the printer and plotter are accessible to the HP as normal RTE logical units. In addition, data files can be transmitted to other peripherals using the file management facility of the data processing system.

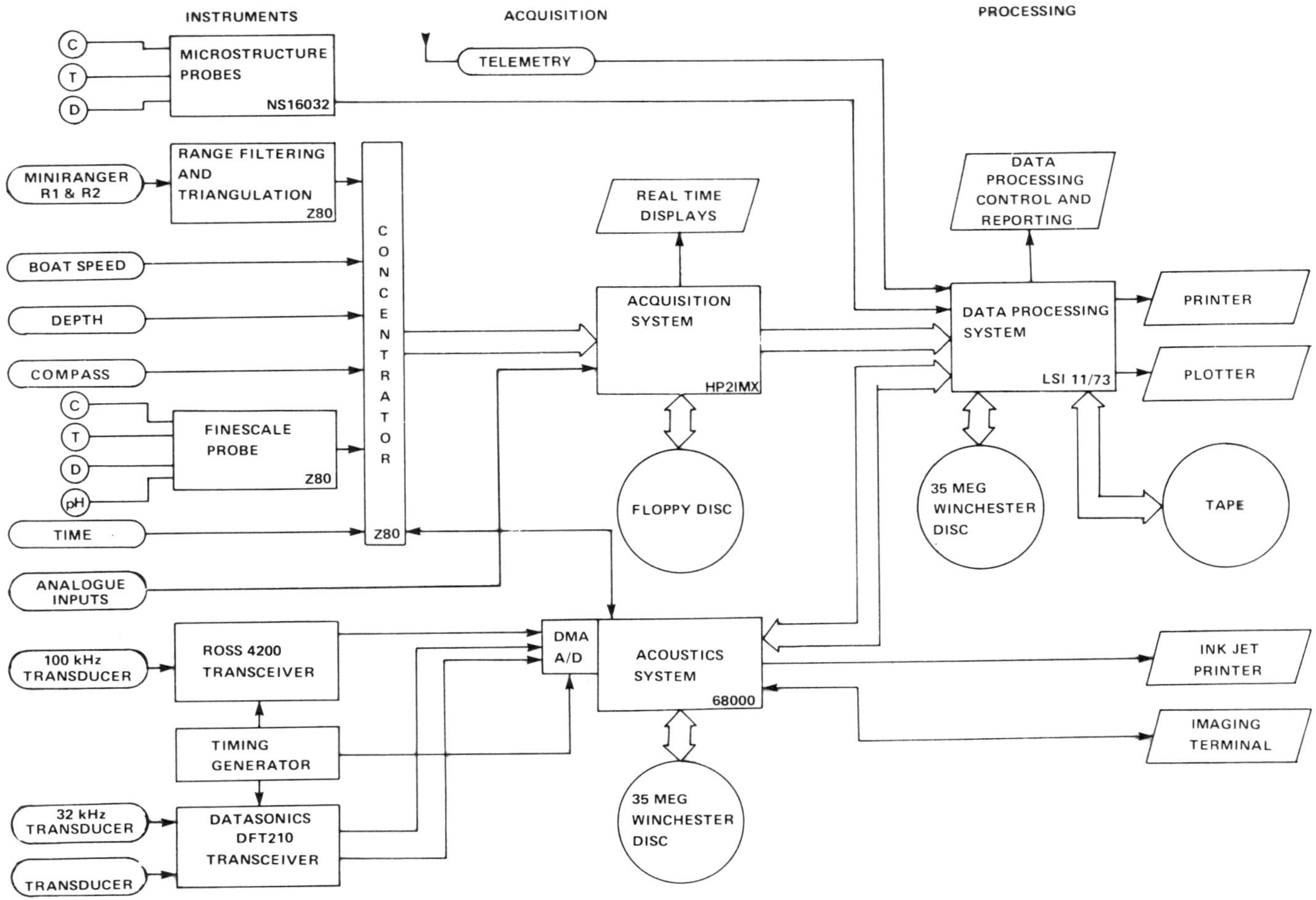

FIGURE 4. Schematic of the Djinnang II data acquisition, acoustic image system, and processing unit.

C. Processing and Numerical Simulation Computer

The requirement here called for a computer able to run many jobs simultaneously. First, it was required that certain processing be done on-line to allow the assessment of the data quality and the success of a particular profile. Quantities such as salinity, Richardson number, dissipation and so forth, all required considerable computing power to be able to make on-line decisions. Similarly it was thought necessary to be able to produce contour plots after completing a particular transect. The requirement of running several numerical simulations as background jobs finally led to the choice of the LSI11/73 computer, a 35-megabyte hard disk and 1 megabyte of memory. The RSX multitasking executive offers good software support for these requirements, and the system has proven to be versatile and flexible. Processing and computing can be carried out via a menu-type operation and all software has been prepared to allow easy batch processing. Again, a dedicated high speed parallel direct memory access bus connects the 11/73 and the HP1000, maximizing availability of the data acquisition system and allowing large scale data processing to be carried out in the field. Another parallel bus interconnects the acoustic imaging system and the data processing system to allow transfer of data sets between these systems. All the parallel buses operate at approximately 1 megabyte per second with very little overhead to any processor.

D. Acoustic Imaging System

The acoustic imaging system has been primarily designed to assist in the optimum deployment of the instruments and provides a real-time picture of the water column, detailing changes to the acoustic impedance arising form turbulence, bubbles, particulate matter, and stratification.[24,25] The images are also stored digitally for later correlation with directly measured parameters, thus providing valuable synoptic information for the interpretation of the data obtained by the direct measurement. Acoustic pulses at three distinct frequencies are transmitted downward and backscatter from within the water column is received, digitized, and recorded. The received data are tagged with position and time information, passed through a two-dimensional filtering algorithm, mapped into a color space, and displayed on a color imaging terminal.

A general description of the system is given in Chapman.[26] Briefly, the front end of the acoustic system uses two commercially available acoustic transceivers. A 100 kHz Ross model 4200 unit transmits 1 kW pulses within a 3.5° beam width. A dual channel Datasonics DFT210 generates up to 4 kW pulses at 32 kHz and 300 kHz with 15° and 2° beams, respectively. All three receivers have input sensitivities of 0.5 μV referred to 50 ohm. Two channels have calibrated gain and time varying gain to permit target strength estimation in association with calibrated transducers. Output power, pulse rate, pulse width, and receiver gain are adjustable on all channels. The receiver output of all channels is digitised at 10 kHz for a burst of 512 samples synchronized to the transmit pulse, yielding 7 cm resolution over 38 m of the water column. For deeper water work, the range may be increased at the expense of vertical resolution by reducing the receiver bandwidth and sample rate by a factor of 2. The three acoustic transducers are mounted on a retractable boom which is suspended at a depth of approximately 0.5 m to the side of the vessel. The transducer pod is housed in a streamlined plastic enclosure which fills with water on submersion, minimising turbulence at the transducer.

VI. DATA ACQUISITION

Data from the primary instruments are combined into a uniform data format by the Z80-based concentrator. Packets of data are then sent to the HP1000 acquisition computer where it is recorded on flexible disk at the incoming data rate. Under normal conditions data are automatically transferred to the data processing system at the completion of data acquisition, and routine processing begins.

A. File Structure

The acquisition program defines 24 physical input channels from which the operator can select those which he wishes to record by using an appropriate menu option or configuration file. The operator must also nominate one channel as a conditional channel and the sample interval. Again, a menu option exists for setting or changing these parameters. Typically, the conditional channel will be time or depth. All data are stored as 32-bit floating point numbers to provide sufficient resolution for all sensors and to allow uniformity in processing.

Associated with each channel is a function number and a set of four calibration coefficients. These are stored in the configuration file and allow the system to automatically calibrate the data before storing or sending it to the processing system. The only parameter specified by the operator is the location name for the station at which the data set is being collected and any notes he may wish to add to the header of the data file. In this way, every parameter associated with the data acquisition process, either introduced manually or retrieved from the configuration file, is recorded in a header at the start of each data file.

Data acquisition has been unified throughout the Djinnang II system, the *in situ* loggers, the PDAS, and the Hydraulics Laboratory data processing system. Many PDAS units exist that service both field and laboratory needs, such as the calibration facility, and all are compatible with the 24-channel format. The floating point representation varies between these machines, but conversion is an automatic function of the transfer programs so that the floating point numbers are always appropriately interpreted, and high level data structures, such as the file header format, are independent of the machine.

B. Configuration File

The configuration files have exactly the same file header format as data files but contain no data. The purpose of a configuration file is to define sampling parameters for the acquisition system in advance of the field trip. Typically, a conditional channel, sample rate/interval, channels to be sampled, and channel labels are stored in the configuration file. Each data file produced refers back to the configuration file used to set up the acquisition system. A configuration file may also contain coefficients for acquisition time calibration of sensors.

C. Calibration File

Instruments with relatively unstable characteristics which must be routinely calibrated (e.g., meteorological station, microstructure probe, etc.) have a calibration file associated with them. This is a text file containing all information relevant to the calibration including extensive notes, calibration residuals, and spot-check values. It also contains the polynomial coefficients in a well-defined format so that they can be read by the acquisition program and used to convert raw voltages to true readings.

VII. SIGNAL ENHANCEMENT

A. Sensor Response Enhancement

Sensors of good stability such as the Sea-Bird temperature and conductivity probes, or the Orion pH sensor usually achieve the stability by sacrificing speed and response. However, the response time may be enhanced digitally by up to a factor of 10, provided the transfer function of the sensor is known. Further, apart from improving the response time of individual sensors, it is very important in both lake and oceanic work[27] to ensure that the response time of two sensors are carefully matched if derived quantities, such as salinity, density or total CO_2 are to be computed without introducing artificial spiking in the derived quantity.

The method used to match the response of the SBE-3 temperature sensor and the SBE-4 conductivity cell is described in Fozdar et al.[28] The technique uses a recursive filter in the time domain which allows direct calculation of salinity and density and thus offers a significant computational advantage over frequency domain methods. Using this method, the useful band width of the SBE-3 temperature sensor has been improved by a factor of 7, with acceptable degradation of the signal-to-noise ratio. In this way, a very close match is obtained between the pump assisted conductivity cell (SBE-4) and the enhanced temperature signal. This removed spikes in the profile of the calculated salinity and density and yet yields an equivalent first-order sensor time constant of about 20 msec.

In the computation of total CO_2, the pH response is sharpened in an equivalent fashion and combined with a digitally filtered and slowed temperature and salinity value. The improvement in pH sensor response is again considerable, decreasing the system response time constant from about 250 msec to 70 msec and producing a spike free total CO_2 profile.

B. Image Processing System

The image processing system provides real-time two-dimensional signal processing capabilities for the enhancement of the acoustic images. The system takes advantage of the inherent asymmetry between the vertical and horizontal axes brought about by the stratification, by implementing a separable two-dimensional filter. This involves the cascading of a high-pass edge enhancement vertical filter with a low-pass horizontal filter to highlight the substantially horizontal structures often found in a stratified flow. In practice, the vertical filter is implemented using a short (5 point) nonrecursive filter while the low-pass characteristics required in the horizontal direction are achieved using a recursive filter so that low spatial cut-off frequencies can be realized in real time. The effective length of the horizontal filter can be varied to match the boat speed and the scales of horizontal variability.

Currently, while all three channels are recorded continuously they are filtered and displayed singly, with the ability to switch between them instantaneously. The filtered data are mapped into color space through a look-up table within the display terminal. This look-up table can be altered at any time in accordance with various algorithms. One algorithm is designed to utilize the full dynamic range of the eye by maximizing the number of distinguishable colors. In practice, this mapping is only used when the signal-to-noise ratio of the image is high enough to compare with the dynamic range of the eye. Another algorithm is designed to produce only colors of high saturation. A nonlinear transformation is also available within the color table generation algorithm to allow localized contrast stretching or compression. This feature allows preferential enhancement in areas of special interest.

VIII. DATA PROCESSING

The data processing may be viewed in all instances equivalent to the data acquisition; the processing computer is used to map a raw data file into a file of processed data or a set of processed files. In this way, files are created in an identical fashion to the acquisition of raw files and so may be viewed as newly created or acquired data files. Viewing processing as such a mapping has greatly simplified data processing and data management; files and sets of files may be mapped in succession, retaining complete universality of all processing and output programs.

A special file-naming and batch-processing system was developed which enables orderly continual enlargement of the system to meet the constantly changing needs of the scientist. The system works by naming the original file with a prefix specifying the data source and the time of acquisition. The prefix is updated at each processing stage, and in addition, the time the new file was created is added to the new file header as a file code. In this way, the identifier of the original file is retained, but the file code uniquely specifies the time of creation in an identical fashion to the original file name. The data flow is schematically indicated in Figure 5.

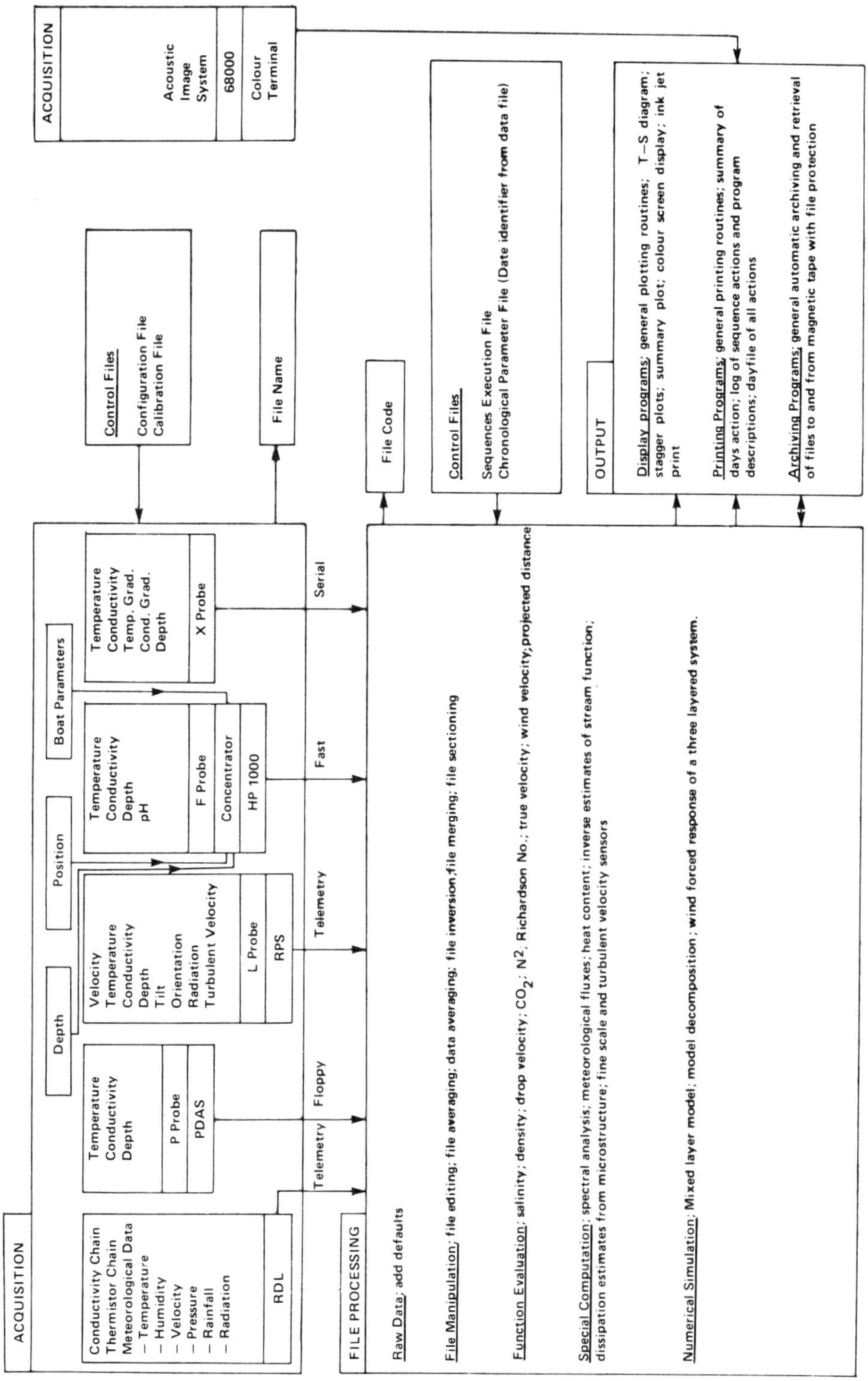

FIGURE 5. Schematic of data flow on the Djinnang II.

A provision has also been included for each action of the process to be written to a history file. Attached to this file is the file name, the file code, and the version number of all the programs used to derive the new processed file. Similarly, any plot produced by the system is identified with both the file name and the file code. In this way, all processing and output is uniquely coded.

The processing is controlled via a menu arrangement either manually or through a predetermined batch sequence. Sequence files may be set up which control the processing from start to finish. The sequence file name may be attached to a file name or a set of files either at the time of acquisition on the HP system or later through the processing system. Typical processing such as computing dissipation from raw data files are routinely done with predetermined sequences. The sequence determines the action of the programs; the additional constants such as filters, conversion coefficients, etc. are all stored in an inventory or chronological coefficient file. As sensor responses change or as the program requirements change, the coefficient file is updated. The user is able to specify which set of coefficients a sequence utilizes by specifying the identifier; the file name is used as a default identifier. In this way the software is secured against obsolescence. The series of fast links make data flow very flexible, allowing quick overlays of contour maps from CTD information and an acoustic image.

Some sequences often take a long time to complete due to the extensive computation involved. Pauses may be inserted at arbitrary points in a sequence causing the batch process to stop and await a restart. The restart may be carried out manually in totality or selectively, or the system may be set to periodically restart all or selected sequences. This feature allows rapid throughput of diagnostic information for field use during an experiment; restarting the sequence overnight allows completion of processing by the next morning, giving the experimentalist a final processed data set before commencing the next phase of his program.

The implementation of pauses, however, quickly led to the accumulation of semicompleted tasks filling the 35 megabyte disk on the 1173. In order to overcome this congestion problem, an automatic archiving program has been written. This program automatically shifts completed files or files awaiting a sequence implementation onto magnetic tape and deletes these from the disk. At the same time the system keeps a directory of the moved files on the disk. Retrieval from tape is thus automatically achieved by the computer, enlarging the effective storage to about one day's operation.

IX. HYBRID MODELING

Field work in lakes and coastal areas is complicated, as explained above, by the great diversity of length scales. This prevents exhaustive detailing of large areas. On the other hand, the basin scale motions may often introduce a Langrangian drift in the area of interest which can sweep the water out of the monitoring area. Such advective motions are often masked by the local processes, yet it is essential for the success of any project to differentiate clearly between advective changes and changes due to local mixing; it is important to retain the correct focus throughout the life of a particular mechanism. For instance, when studying mixed layer deepening it is essential that sufficient data be gathered on the right scale and in the right location to allow the separation of deepening of the mixed layer by vertical entrainment, tilting of the base of the mixed layer, and apparent deepening due to advective changes.

It has been found[2] that this separation of motions presents great difficulties in the field. However, experience has shown that the 11/73 may be used for direct comparisons of numerical simulated results with field data. A suite of programs is under development to allow field initialization of numerical models of particular processes under study. The output from these models is compatible in file structure, and output comparisons can be introduced into a sequence. A mixed layer model WML6[29] has been successfully used in this way. The model, a one-dimensional vertical mixed layer model, may be initialized with a field data profile. The meteorological data received by telemetry is used to compute the predicted water profile characteristics at the current time. Comparison with field data allows clear separation of advective and tilting effects from vertical entrainment, as well as focusing attention on any billow activity.

Future program development will include gravitational overflow predictions,[30] reservoir simulation prediction through the model DYRESM,[31] and a general wind driven circulation model. These models will be used together with drogue information to assist the field scientist to focus on particular dynamical regimes in the water column.

X. EXAMPLE

It is evident from the above discussion that the flexibility of the system is such that an almost continuous spectrum of configurations both for data acquisition and data processing are possible. Space here does not permit more than a very simple illustration of the system.

Figure 6 shows a sample output from a cast with the F probe taken on Julian day 265 in 1984 in Koombana Bay south of Perth in Western Australia. The cast was one of many taken as part of a field investigation of the dynamics of a gravitational overflow. The aims of the investigation were to measure both the mean and turbulent properties of a gravitational overflow[30] in order to determine the propagation speed, the rate of collapse of the frontal roller, the loss of energy by internal wave generation, and the overall efficiency of the overflow.

The raw variables are shown in Figures 6a to c, the corrected and sharpened variables in Figure 6d to f, and some derived functions in Figure 6g to i. The effective sharpening is quite marked and yields a spatial resolution of about 0.02 m, with a vehicle drop velocity of 1 m sec^{-1}. In Figure 6f, static instabilities are observed at 1.1 and 1.6 m. These instabilities are reflected in the displacement scale shown in Figure 6g where the root mean squared displacement l (bin size 0.4 m) is plotted. The diffusion coefficient (Figure 6h) was estimated from the expression Nl^2, where N is the buoyancy frequency. Particularly interesting is the patchy nature of the estimated vertical diffusion coefficient. The total CO_2 content can be used to study the net community production and consumption of dissolved inorganic carbon throughout the water column.[32]

XI. SUMMARY

A flexible data acquisition and data processing system has been described. This system was designed specifically to study mixing in a stratified geophysical flow found in lakes and coastal regions. By combining a series of new measurement instruments with remote sensing of the water column and a hierarchy of microcomputers, an extremely effective facility has been created. The hardware and software have been designed to allow expansion without rendering old data files incompatible. The facility has already found application in investigations of mixed layer dynamics, inflows and intrusions in lakes, gravitational overflows in coastal seas, and boundary mixing.

ACKNOWLEDGMENTS

David Pullin, Geoff Prince, John Brubaker, and John Patterson contributed to the software development. Geoff Bishop, Ken Bentley, Les Lidbury, and Brian Sambell are responsible for the maintenance of the facility, and Farhad Fozdar, Geoff Carter, Arthur Hyde, and Tony Jenkinson designed and built much of the electronic hardware. Les Reid is the technical

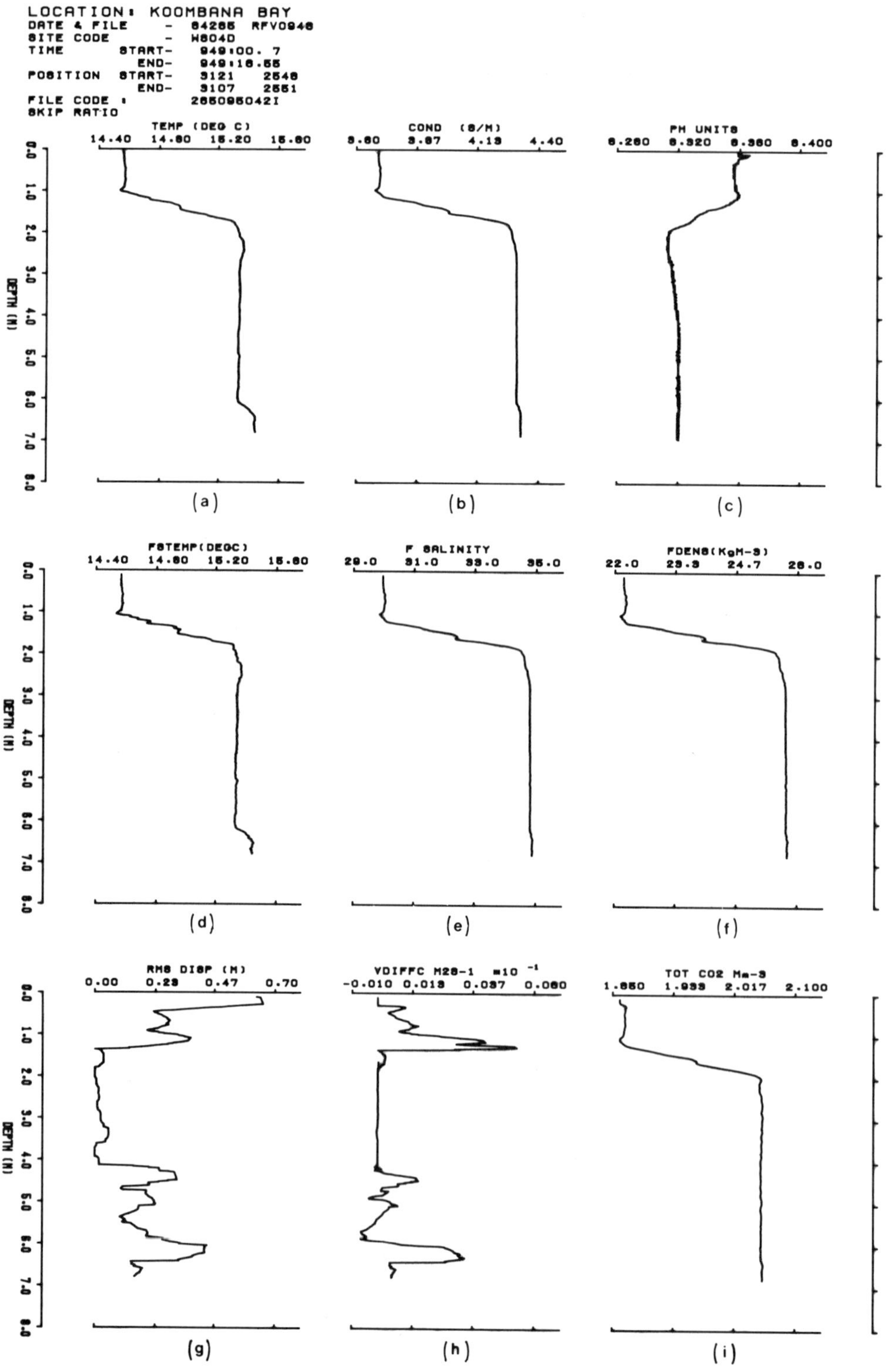

FIGURE 6. Example of a CTD data set from the F probe. (a) raw temperature, (b) raw conductivity, (c) raw pH all sampled at 50 Hz, (d), (e), (f) enhanced signals, (g) root mean square displacement, (h) estimated vertical diffusion coefficients, (i) total CO_2 content of water.

administrator. The facility is financially supported by the Australian Research Grants Scheme, the Australian Water Resources Council, the Commonwealth Research Centres of Excellence scheme, the University of Western Australia, and the Centre for Water Research. The first author was visiting the California Institute of Technology under the Fairchild Fellowship Program, their assistance is gratefully acknowledged.

REFERENCES

1. **Fischer, H. B., List, E. J., Koh, R. C. Y., Imberger, J. and Brooks, N. H.,** *Mixing in Inland and Coastal Waters,* Academic Press, New York, 1979.
2. **Imberger, J.,** The diurnal mixed layer, *Limnol. Oceanogr.,* 30, 737, 1985.
3. **Garvine, R. W. and Munk, J. D.,** Frontal structure of a river plume, *J. Geophys. Res.,* 79, 2251, 1977.
4. **Thorpe, S. H. and Hall, A. J.,** Observations of the thermal structure of Langmuir circulation, *J. Fluid Mech.,* 114, 237, 1982.
5. **Csanady, G. T.,** *Circulation in the Coastal Ocean,* D. Reidel, Dordrecht, The Netherlands, 1982.
6. **LeBlond, P. H. and Mysak, L. A.,** *Waves in the Ocean,* Elsevier, Amsterdam, 1978.
7. **Imberger, J.,** Thermal characteristics of standing waters: An illustration of dynamical processes, in *Perspectives in Southern Hemisphere Limnology,* Vol. 125, Davies, B. R. and Walmsley, R. D., Eds., Hydrobiologia, 1985, 7.
8. **Corcos, G. M. and Sherman, F. S.,** Vorticity concentrations and the dynamics of unstable shear layers, *J. Fluid Mech.,* 73, 241, 1976.
9. **Imberger, J. and Hamblin, P.,** Dynamics of lakes, reservoirs and cooling ponds, *Annu. Rev. Fluid Mech.,* 14, 153, 1982.
10. **Veronis, G.,** Inverse methods for ocean circulations, unpublished manuscript, 1987.
11. **Chips Brothers,** Remote data logger, Environmental Dynamics Report ED-82-017, University of Western Australia, Perth, 1982.
12. **Chips Brothers,** A low cost logger for recording borehole level or rainfall and windspeed, Environmental Dynamics Report ED-83-057, University of Western Australia, Perth, 1983.
13. **Fozdar, F. M.,** Remote profiling station technical report, Environmental Dynamics Report ED-85-096, University of Western Australia, Perth, 1985.
14. **Scolaro, G. and Luketina, D.,** Laboratory data acquisition system technical report, Environmental Dynamics Report ED-84-078, University of Western Australia, Perth, 1984.
15. **Fozdar, F. M.,** An introduction to the mixed layer probe, *Measurement,* 2, 103, 1984.
16. **Fozdar, F. M.,** Mixed layer probe technical report, Environmental Dynamics Report ED-83-039, University of Western Australia, Perth, 1983.
17. **Caldwell, D. R., Dillon, T. M., Brubaker, J. M., Newberger, P. A. and Paulson, C. A.,** The scaling of vertical temperature gradient spectra, *J. Geophys. Res.,* 85, 1917, 1980.
18. **Dillon, T. M.,** Vertical overturns: A comparison of Thorpe and Ozmidov length scales, *J. Geophys. Res.,* 87, 9601, 1982.
19. **Shay, T. J. and Gregg, M. C.,** Turbulence in an oceanic convective mixed layer, *Nature (London),* 310, 282, 1984.
20. **Washburn, L. and Gibson, C. H.,** Measurements of oceanic temperature microstructure using a small conductivity sensor, *J. Geophys. Res.,* 87, 4230, 1982.
21. **Caldwell, D. R. and Dillon, T. M.,** An oceanic microstructure measuring system, School of Oceanography Ref. 81-10, Oregon State University, Corvallis, 1981.
22. **Carter, G. D., Scolaro, G., Elford, D. and Luketina, D. A.,** Vertically rising microstructure profiler, Environmental Dynamics Report ED-85-097, University of Western Australia, Perth, 1985.
23. **Chips Brothers,** Djinnang II equipment technical report, Environmental Dynamics Report ED-83-059, University of Western Australia, Perth, 1983.
24. **Thorpe, S. A. and Brubaker, J. M.,** Observations of sound reflection by temperature microstructure, *Limnol. Oceanogr.,* 28, 601, 1983.
25. **Thorpe, S. A.,** On the determination of $K\nu$ in the near-surface ocean from acoustic measurements of bubbles, *J. Phys. Oceanogr.,* 14, 855, 1984.
26. **Chapman, R. H.,** Acoustic imaging system for turbulent microstructure, Environmental Dynamics Report ED-85-095, University of Western Australia, Perth, 1985.
27. **Gregg, M. C. and Hess, W. C.,** Dynamic response calibration of the Sea-Bird temperature and conductivity probes, *J. Atmos. Oceanic Technol.,* submitted, 1987.

28. **Fozdar, F. M., Parker, G. J., and Imberger, J.,** Matching temperature and conductivity sensor response characteristics, *J. Phys. Oceanogr.,* 15, 1557, 1985.
29. **Spigel, R. H., Imberger, J. and Rayner, K. N.,** Modelling the diurnal mixed layer, *Limnol. Oceanogr.,* 31, 533, 1986.
30. **Imberger, J.,** Tidal jet frontogenesis, *Mech. Eng. Trans.,* I. E. Aust., ME8, 171, 1983.
31. **Imberger, J. and Patterson, J. C.,** A dynamic reservoir simulation model-DYRESM:5, in *Transport Models for Inland and Coastal Waters,* Fischer, H. B., Ed., Academic Press, New York, 1981, 310.
32. **Skirrow, G.,** The dissolved gases — carbon dioxide, in *Chemical Oceanography,* Vol. 2, Riley, J. P. and Skirrow, G., Eds., Academic Press, New York, 1975, 1.

INDEX